MECHANICAL RELIABILITY IN THE PROCESS INDUSTRIES

Seminar Planning Panel

T. R. Moss, MPhil, CEng, FIMechE, FIQA
RM Consultants
Didcot

R. J. Aird, BSc, CEng, MIMechE
University of Technology
Loughborough

C. J. Austin, CEng, MRAeS, MIGasE
British Gas Corporation
London

A. G. Cannon, BSc(Hons), FInstP
National Centre of Systems Reliability UKAEA
Warrington

A. O. F. Venton, CEng, MIMechE, FIMarE, FSRS, MBIM
RM Consultants Ltd
Didcot

A. N. Kinkead, MSc, CEng, MIMechE
National Centre of Systems Reliability UKAEA
Warrington

R. F. de la Mare, BSc, MBA, PhD, CEng, FIChemE, MBIM
University of Bradford
Bradford

A. K. Gardner, BSc, CEng, FIMechE, MASME
Imperial College
London

MECHANICAL RELIABILITY IN THE PROCESS INDUSTRIES

Papers presented at a seminar jointly organized by the Process Industries Division of The Institution of Mechanical Engineers and the National Centre of Systems Reliability, Mechanical Reliability Research Group

Published by
Mechanical Engineering Publications Limited for
The Institution of Mechanical Engineers
LONDON

First published 1984

ISBN 0 85298 552 5

Printed by Waveney Print Services Ltd, Beccles, Suffolk

CONTENTS

Design for reliability 1
A. D. S. Carter, P. Martin, and A. N. Kinkead

Component based prediction for mechanical reliability 11
A. O. F. Venton and T. R. Moss

The influence of some tribological problems in mechanical reliability 21
R. S. Sayles, M. N. Webster, and P. B. MacPherson

Mechanical value reliability 31
R. J. Aird and T. R. Moss

Pump reliability in relation to economised maintenance 37
R. F. de la Mare

Implementation of a reliability programme on a new gas compression station – a case history 45
T. P. Littlejohns and M. Kelly

Factors affecting reliability specifications for process plants 53
D. J. Sherwin

The Institution of Mechanical Engineers

The primary purpose of the 76,000-member Institution of Mechanical Engineers, formed in 1847, has always been and remains the promotion of standards of excellence in British mechanical engineering and a high level of professional development, competence and conduct among aspiring and practising members. Membership of IMechE is highly regarded by employers, both within the UK and overseas, who recognise that its carefully monitored academic training and responsibility standards are second to none. Indeed they offer incontrovertible evidence of a sound formation and continuing development in career progression.

In pursuit of its aim of attracting suitably qualified youngsters into the profession — in adequate numbers to meet the country's future needs — and of assisting established Chartered Mechanical Engineers to update their knowledge of technological developments — in areas such as CADCAM, robotics and FMS, for example — the IMechE offers a comprehensive range of services and activities. Among these, to name but a few, are symposia, courses, conferences, lectures, competitions, surveys, publications, awards and prizes. A Library containing 150,000 books and periodicals and an Information Service which uses a computer terminal linked to databases in Europe and the USA are among the facilities provided by the Institution.

If you wish to know more about the membership requirements or about the Institution's activities listed above — or have a friend or relative who might be interested — telephone or write to IMechE in the first instance and ask for a copy of our colour 'at a glance' leaflet. This provides fuller details and the contact points — both at the London HQ and IMechE's Bury St Edmunds office — for various aspects of the organisation's operation. Specifically it contains a tear-off slip through which more information on any of the membership grades (Student, Graduate, Associate Member, Member and Fellow) may be obtained.

Corporate members of the Institution are able to use the coveted letters 'CEng, MIMechE' or 'CEng, FIMechE' after their name, designations instantly recognised by, and highly acceptable to, employers in the field of engineering. There is no way other than by membership through which they can be obtained!

Design for reliability

A D S CARTER, P MARTIN and A N KINKEAD

SYNOPSIS

Mechanical design involves much empiricism and is concerned with many different aspects of equipment and system behaviour in addition to reliability. Totally new methods of design, concentrating on one aspect, will therefore usually be rejected by established designers. In this paper the design requirements to achieve specified reliability for a particular life have been expressed in terms that mechanical design engineers, in addition to reliability engineers, can appreciate and relate to the conventional design process. Because the subject is so extensive, the paper concentrates on indicating sufficient detail in two important areas to provide a working design document.

Stress-rupture mechanisms of failure are considered initially. For these mechanisms standard techniques to achieve reliability are usually based on the safety margin concept. Existing criteria are compared and re-interpreted in terms of conventional factors of safety which can be accommodated by traditional design methods. The factors dictating reliability are expressed in terms of physical parameters which can be directly interpreted. The dependence of the factor of safety on these reliability parameters is then the only departure from the traditional mechanical design method.

A specified reliability relative to stress-rupture mechanisms does not safeguard against the effects of material degradation. The range of degradation mechanisms is exceedingly large, but one important mechanism, fatigue, has been used as an illustration in this paper. The Miner summation is widely used for initial estimates in traditional deterministic design. A stochastic design approach allows for the distributional nature of fatigue life at various stress levels and predicts a cumulative usage factor and reliability for a specified life. The relationship between the two approaches is shown in the paper.

Thus it is shown in two important cases that design for reliability does not involve a fundamental departure from traditional techniques, but could considerably enhance their value because reliability is positively incorporated.

1. INTRODUCTION

It is often claimed, with some justification, that design is the key to mechanical reliability: it is too expensive to modify the production line of a mass produced item - it is applicable even when the quantities are relatively small - and one clearly cannot replace the expensive "one off" item. Usually one is "stuck with" a design for better or worse once manufacture has begun. Much effort has gone into optimising the existing product through maintenance schedules, method of operation, life cycle costs and so on, but the optimum of a second rate system is itself second rate. Yet the extensive background knowledge on reliability has rarely "got across" to designers, who by and large ignore the vast pool of information that is now becoming available to them. Possibly this can be understood. There is a lot more to design besides reliability: obviously of prime importance are the functional aspects, including level of performance and running costs, size, weight and so on - particularly in relation to market competition and the permissible degree of innovation - then there are manufacturing aspects including feasibility within the resources of an organisation, standardisation, and costs; there are such aspects as sub contracting and the availability of bought-in parts, and there are often even aesthetic features to be considered. Much of this is empirical and designers do not take kindly to new methods of design that could possibly influence this empiricism in an unknown manner. The way ahead therefore demands that reliability workers talk to designers in designer's language, even if this means that some scientific exactitude is lost in the process. The object of this paper is primarily to show that it can be done. The complete field is so wide that concentration on limited topics is necessary if worthwhile information is to be contained within a single paper, for the second objective of the paper is to provide working methods that can be used in the design process for the topics covered.

Two aspects have been chosen for treatment. These are thought to encompass a wide range of general mechanical engineering design problems.

The first concerns design against failure by a stress-rupture mechanism, i.e. one in which past history plays no part and it is the instantaneous load which exceeds the invariant strength capacity of an item at failure.

Even if no failures ever arise in this mode an item may still fail through a wear out mechanism, i.e. one in which past history is the controlling mechanism through a reduction of the strength capacity by loads which are not high enough to cause stress-rupture failure. The most important of these mechanisms is probably fatigue. Thus fatigue forms the second major topic discussed in this paper. It must be recognised that other failure mechanisms can arise, but it is believed that if design in these two areas can be based on sound reliability techniques a major step forward will have been taken.

2 NOTATION

A	area
c	intercept on y axis in linearised approximation
f(Ni)	probability density function of cycles to failure Ni
K	constant in max load/min strength method
k	constant in difference distribution method
$\underline{L}$	load
$\bar{L}$	mean load
LR	loading roughness
L(s)	probability density function of load
m	slope of linearised approximation
N	cycles of fluctuating stress
Ne	damage equivalent of N cycles
$\underline{Ni}$	i cycles of fluctuating stress
$\bar{Ni}$	mean of distributed values of Ni
n	number of load applications
P	material property
R	reliability
R(N)	reliability-probability of survival for N cycles
Ri(N)	reliability-probability of survival for N cycles after ith stress level
r	stress ratio alternating/mean stress
Sa	fluctuating test stress amplitude
Se	endurance limit
$\underline{Sf}$	failure governing strength
$\bar{Sf}$	mean failure governing stress
S_{fei}	equivalent fatigue strength for Ni cycles
$\bar{S}_{fei}$	mean fatigue strength at Ni cycles
SM	safety margin
Sm	mean test stress
$\underline{Su}$	ultimate strength
$\bar{Su}$	mean ultimate strength
s	stress
sa	applied stress amplitude
$\underline{sf}$	failure governing stress
$\bar{sf}$	mean failure governing stress
s_{fe}	mean equivalent load stress
sm	applied mean stress
V	coefficient of variation of load (loading factor)
V_{ni}	coefficient of variation of cycles to failure at load stress u
Z	standardised variable of normal distribution
γ	coefficient of variation of material property
$\underline{\delta}$	strength
$\bar{\delta}$	mean strength
$\delta(s)$	probability density function of strength
σ_L	standard deviation of load
σ_{Ni}	standard deviation of distributed values of Ni
σ_P	standard deviation of material property
σ_{Sf}	standard deviation of failure governing strength
σ_{sa}	standard deviation of applied alternating stress
σ_{sf}	standard deviation of failure governing stress
σ_δ	standard deviation of strength
Φ	parameter of standardised normal distribution

3 DESIGN TO CONTROL STRESS-RUPTURE FAILURES

3.1 Concept from reliability theory

There are several design techniques based on general reliability theory which purport to control stress rupture failures at a desired level. They are all based on the assumption that loads and strengths are distributed in some form, as illustrated in Fig. 1. The object is then to separate the distributions, i.e. to achieve a margin, so that only an acceptable level of interaction, and hence failures, can take place between the two distributions.

One of the simplest assumptions to make is that both load and strength are normally distributed. It then follows that their differences will also be normally distributed with a mean of $(\bar{\delta} - \bar{L})$ and a standard deviation of $\sqrt{(\sigma_\delta^2 + \sigma_L^2)}$; where $\bar{\delta}$ and $\bar{L}$ are the mean strength and mean load respectively, and σ_δ and σ_L are the standard deviations of the strength and load respectively. Failures will occur when the load exceeds the strength, i.e. when $(\bar{\delta} - \bar{L})$ is negative. This is illustrated in Fig. 2 in which the probable failures are represented by the shaded area. In order to control their magnitude it is necessary that the mean of the difference distribution, i.e. $(\bar{\delta} - \bar{L})$, be an appropriate number of standard deviations above zero. Hence

$$\bar{\delta} - \bar{L} = k\sqrt{(\sigma_\delta^2 + \sigma_L^2)} \qquad (1)$$

where k is a constant defining the proportion of acceptable failures. Equation (1) can be re-written as

$$\text{Safety Margin (SM)} = \frac{\bar{\delta} - \bar{L}}{\sqrt{(\sigma_\delta^2 + \sigma_L^2)}} = k \qquad (2)$$

This approach thus results in the safety margin being the unique parameter defining reliability, and the parameter to be used in design. The value of k may be deduced directly from the normal distribution, or may be based on experience.

An alternative method is to assume that all loads and strengths on which the design is to be based are contained within 3K standard deviations from the mean, where K is some constant, derived from experience, close to unity. Design is then based on the long standing criterion that the minimum strength must be greater than the maximum load. Hence

$$\bar{\delta} - 3K\sigma_\delta > \bar{L} + 3K\sigma_L \qquad (3)$$

$$\text{or} \quad (\bar{\delta} - \bar{L}) > 3K\,(\sigma_\delta + \sigma_L) \qquad (4)$$

It is to be noted that the mean load and mean strength are separated by a fixed number of standard deviations of load plus the same number (though this could be changed) of standard deviations of strength, as contrasted with a fixed number of standard deviations of the difference in the first approach. Thus in the second method, both standard deviations play individual roles. Now loading roughness is a measure of the load standard deviation expressed in non-dimension terms. Thus

$$\text{Loading Roughness (LR)} = \frac{\sigma_L}{\sqrt{(\sigma_\delta^2 + \sigma_L^2)}} \qquad (5)$$

It can then be shown that the minimum value of the safety margin required to meet the condition (4) is given by

$$(SM) = 3K \left\{ (LR) + \sqrt{\left[1 - (LR)^2 \right]} \right\} \quad (6)$$

so that the safety margin is not a constant as in the first approach described. Differentiating (6) and equating to zero shows that the maximum value of the required safety margin is 4.424 K at a loading roughness of 0.707. The variation of safety margin with loading roughness over the whole range is presented in Fig. 3. In no way can the two approaches be regarded as equivalent: the difference in the required safety margin over the loading roughness range in the second method is too great to neglect.

A major criticism of both the methods described above is that neither takes cognizance of repeated loads. Failure is assumed to occur because <u>the</u> load is greater than <u>the</u> strength, though both are taken from statistical distributions. One consequence of this is that both methods design for <u>a</u> reliability which is independent of time. This may be true for ideally smooth loading, but contrasts markedly with the well established negative exponential distribution, which represents many real data. As an alternative to the above approaches the senior author has evaluated the reliability of an item subject to repeated loading from arbitrary distributions by means of the equation

$$R = \int_0^\infty \delta(s) \left\{ \int_0^S L(s)\, ds \right\}^n ds \quad (7)$$

where n is the number of load applications. It is demonstrated in (1) that the safety margin can be chosen so that an item is intrinsically reliable, which has been described in that reference as having a failure rate which is immeasurably low and can be taken as zero, even when n is allowed to take high values. The minimum safety margin to achieve intrinsic reliability has also been plotted in Fig. 3 for comparison.

There is a good deal of similarity in the various criteria, though as demonstrated in Fig. 3 there can be considerable numerical differences. Of some importance they all use the safety margin as a parameter for design. However, this is often difficult to use in practice, and means little to the standard designer who is unable to relate it to physical circumstances and strongly prefers to use the well tried and established factor of safety. Our task must therefore be to relate the safety margin to the factor of safety.

3.2 Relationship between Safety Margin and Factor of Safety

Clearly no estimate of the working strength can be made without a knowledge of the load being applied to the item. It will consequently be assumed that information is available which will enable both the mean load, $\bar{L}$, and its standard deviation, σ_L, to be deduced.

The strength of an item will be related to the properties of the material so that

$$\frac{\sigma_\delta}{\bar{\delta}} = \frac{\sigma_P}{\bar{P}} = \gamma \quad (8)$$

where σ_P and $\bar{P}$ are the standard deviation and mean of the relevant material property, and γ is its coefficient of variation. Thus in application to a simple tensile member, for example

$$\bar{\delta} = A\,\bar{P} \quad (9)$$

$$\sigma_\delta = A\,\sigma_P \quad (10)$$

where A is the cross section area of the member, and P is the proof stress, or yield stress or perhaps the ultimate tensile stress depending on the nature of the design. Similar considerations apply to other loadings.

The design conditions discussed in sub-section 3.1 are best handled by linearising the safety margin - loading roughness relationship. Thus

$$(SM) = m\,(LR) + C \quad (11)$$

or

$$\frac{\bar{\delta} - \bar{L}}{\sqrt{(\sigma_\delta^2 + \sigma_L^2)}} = m \frac{\sigma_L}{\sqrt{(\sigma_\delta^2 + \sigma_L^2)}} + C \quad (12)$$

Letting the coefficient of variation of the load $V = \sigma_L/\bar{L}$ and substituting for $\sigma_\delta = \gamma\bar{\delta}$ from equation (8) into equation (12) gives a quadratic equation in $\bar{\delta}$ which can be solved to give

$$\frac{\bar{\delta}}{\bar{L}} = \frac{\{1+m(V)\}+C\sqrt{\{(V)^2(1-C^2\gamma^2)+\gamma^2\left[1+m(V)\right]^2\}}}{(1 - C^2\gamma^2)} \quad (13)$$

where the positive sign of the square root has been taken as that relevant to this problem. The ratio $\bar{\delta}/\bar{L}$ is, of course, the conventional factor of safety. In some applications it is usual to base the factor on the maximum load rather than the mean but this is a minor correction once the distribution is known.

As illustrated, the full line curve for intrinsic reliability as shown on Fig. 3, has been replaced by the following piece-wise linearisation,

For smooth loading, i.e. for $0 < LR < 0.5$

$$(SM) = 3.8\,(LR) + 3.1 \quad (14)$$

For medium smooth/rough loading, i.e. for $0.5 < LR < 0.8$

$$(SM) = 1.66\,(LR) + 4.16 \quad (15)$$

For rough loading, i.e. for $0.8 < LR < 1.0$

$$(SM) = 5.5 \quad (16)$$

The resulting factors of safety have been plotted against the load coefficient of variation for various values of the material coefficient of variation in Fig. 4. Similar curves could be produced for the other design criteria discussed in sub-section 3.1, or for other load and strength distributions. It is just as valid to use these factors of safety in design, and achieve exactly the same result, as using safety margins directly.

3.3 Comments on design with stress-rupture mechanisms

It has been shown that stochastic design involving stress-rupture mechanisms can be accomplished using conventional factors of safety. It is recommended that design techniques based on reliability theory be expressed in this way, for it enables all existing experience and expertise to be retained and directly weighed against theory, an aspect believed to be of immense value and importance. There is no need for designers to abandon their past experience and the "gut feeling" they bring to design. At the same time the recommended technique rationalises the approach by virtue of its theoretical basis, and lays emphasis on a proper understanding of the contributions made by the loading conditions and the material properties. Although some intuitive anticipation will still be necessary during design it will be the anticipation of physical quantities that can subsequently be measured and checked against assumption, thus ever improving future estimates. This is clearly an advance over the intuitive assessment of a factor of safety, which is an abstract quantity that can never be measured. It is important to note, however, that the two parameters, γ and $V=(\sigma_L/L)$, completely define the factor of safety, and no other factor of safety can logically be applied. Thus the so-called "factor of ignorance" has no place in this logic. If there is any ignorance it applies to the parameters themselves and any such correction should be applied at source where it will be properly weighted in the final result. It is imperative therefore that values assigned to the above parameters be comprehensive.

No difficulty is anticipated in measuring the material property, γ, at least when samples of the finished product are available. This is to be noted, for γ contains more than the basic material properties. It must be based on the material properties of the finished product - not a test piece - and include any modifications introduced by the manufacturing process, by the form of the item (including tolerances), by the finishing process, and so on. Even so no great difficulty is seen in assigning a value to γ.

It is believed that the loading factor $(\sigma_L/\bar{L})$ will present more difficulties, both in the initial assessment and in subsequent measurements. This quantity is still so elusive that experience will be at a premium, though more data is becoming available as its importance is more widely recognised. In this connection it must be recalled that the values on Fig. 4 refer to normally distributed loads and strengths (actually Weibull distributions with a shaping parameter of 3.44).

Finally it must be recorded that this section has only been concerned with the estimation of the working strength to withstand an identified load by a specified geometry when a stress-rupture failure mechanism is anticipated. The identification of the loads is not covered in any way; neither are alternative methods of supporting the load. These remain areas of vital importance in design in which the intuitive approach and experience remain paramount.

4 DESIGN TO CONTROL FATIGUE FAILURES

It is of course now widely recognised that the components of machines, systems and plants can be subjected to cyclic mechanical and thermal stresses which result in progressive material damage during component operation. As a result components can fail at stress levels well below their nominal static strength by a mode of failure known as fatigue. This mode of failure is thought to account for a large proportion of mechanical failures. The proportion is often quoted as lying somewhere between 80% and 90%. The European Communities (2) further underlined the importance of this phenomena in relation to the service life of industrial products by indicating that fatigue failure accounts for 95% of all fracture damage in service operation. Whatever the real figure, it has long been recognised that this failure mode has serious safety and reliability implications for mechanical components. As a result, the phenomenon has been widely investigated and extensive information is now available in the literature explaining many aspects of the physical mechanisms involved as well as providing material data and mathematical models which are intended to provide guidance during the design of mechanical components. Many design methods are intended to avoid fatigue failure in service but do not provide any numerical indication of the reliability of components over their intended life. However, much recent work has been aimed at establishing links between existing design methods and reliability prediction methods suitable for use during design. In this short paper it is hoped that the basis of some of this work can be illustrated.

Clearly only a few aspects can be considered and only methods applicable to the fatigue of uncracked components subjected to high cycle mechanical fatigue stresses are examined. In principle this component category would include many types of component which are machined and where surface imperfections are well controlled, such as bolts, shafts, bearings, gear teeth, etc., as opposed to cracked structures such as bridges, ships, pressure vessels, etc.

4.1 High cycle fatigue of uncracked components

It is now widely appreciated that the fatigue failure of uncracked components involves the nucleation of macro-cracks followed by crack growth to the point where the load can no longer be carried and failure occurs. Since a general deterministic law for crack nucleation time is not available, design to control fatigue failure is normally based on the interpretation of material fatigue test results. In order to relate this procedure to fatigue reliability estimation, a typical testing procedure is now briefly considered.

High cycle fatigue tests on material specimens are usually carried out by cycling the material using either a tension and compression test or a rotating bend test, between an upper stress level and a lower stress level about a zero mean test stress (Sm = 0) to give a test stress amplitude S with the sinusoidal form shown in Fig. 5. When a number of specimens are cycled to failure at different values of stress amplitude S_a, a typical S-N characteristic with the form shown in Fig. 6 then results. This indicates the number of cycles to failure at each test stress amplitude. Typically, a test piece subjected to a stress amplitude S_{fe1} would survive for a life of N_1 cycles. Below a particular test stress amplitude S_e (the endurance limit) some materials and in particular many steels survive for an infinite number of loading cycles.

When it is necessary to determine the strength of the material for a component subjected to a mean stress sm which is not zero, the stress amplitude S_{fe1} must be adjusted to allow for the presence of the mean stress to produce failure at the same number of cycles N_1. A number of empirical relationships have emerged which allow for this problem. For example the well known modified Goodman Diagram, Fig. 7, gives the lower alternating stress amplitude, Sa_1, which combined with a mean stress Sm_1, would give a life of N_1 cycles relative to the ultimate strength of the material, Su, and this can be expressed as

$$Sa_1 = S_{fe1} \left(1 - \frac{Sm_1}{Su}\right) \quad (17)$$

Due to the empirical nature of this relationship, it would be dangerous to work close to the predicted stress levels. An alternative approach is to use the modified Goodman Diagram as a basis for design in the form shown in Fig. 8. In this case the diagram is based on the endurance strength S_e. The Goodman line and the shaded region above it represent a failure region. A component designed to carry a peak mean stress sm_1 and a fluctuating stress of amplitude sa_1 would then lie in a safe region of the diagram. Such stresses could for example represent the stresses on a preloaded bolt after allowing for stress concentration and surface effects, etc. A more conservative design approach is based on using the failure region bounded by the Soderberg Line which is based on the yield point of the material S_{YP} or the less conservative Gerber Line which is defined by the relationship

$$\left(\frac{Sa}{Se}\right)^1 + \left(\frac{Sm}{Su}\right)^2 = 1 \quad (18)$$

Considerable design judgement is required in the sizing of components even after allowing for the usual design stress correction factors in arriving at suitable values for the mean design stress sm_1 and the alternating design stress sa_1 relative to these failure regions.

The Goodman diagram in Fig. 8 may be used to provide a basis for defining a safety factor. The failure governing strength Sf may be taken as

$$Sf = (Sa_1^{\,2} + Sm_1^{\,2})^{\frac{1}{2}} \quad (19)$$

and the failure governing stress sf may be taken as

$$sf = (sa_1^{\,2} + sm_1^{\,2})^{\frac{1}{2}} \quad (20)$$

where the stress ratio is

$$r = \tan\alpha = sa_1/sm_1 \quad (21)$$

and the Safety Factor S.F. may then be regarded as

$$S.F. = Sf/sf \quad (22)$$

However, this safety factor alone does not readily allow for the well known fact that there is often considerable scatter in fatigue test data. The reliability approach may be regarded as an attempt to allow for this scatter in making design decisions on component life.

4.2 Fatigue Reliability of Uncracked Components

In considering the reliability analysis of mechanical components and systems, Kececioglu (3) has used the concepts of failure governing strength and failure governing stress as considered in equations (19) and (20) and has presented fatigue test results for a number of materials which demonstrate that it is more realistic to consider the results to be scattered about a mean line as opposed to representing fatigue test results in the form of the unique line shown in Fig. 6. Test results can then be represented by a number of cycles to failure and stress to failure distributions as shown in Fig. 9.

For illustrative purposes, the determination of the reliability for a life of N_1 cycles of a tensile member subjected to a fluctuating stress sa_1 with a standard deviation σ_{sa1} and a mean stress sm_1, is considered. Kececioglu and Chester (4) have demonstrated that for axial tensile stresses, the probability density functions of cycles to failure can be regarded as normally distributed and the Gerber line adequately represents the mean failure boundary.

The following steps involved in the determination of the reliability of the tensile member with a fluctuating load for a life of N_1 cycles are based essentially on the approach of Kececioglu (3).

(a) sa_1 and sm_1 are known hence

$$\text{Stress ratio } r = \frac{sa_1}{sm_1} = \frac{Sa_1}{Sm_1} \quad (23)$$

(b) The alternating strength $\bar{S}_{fe1}$ is obtained from the S-N diagram at N_1 cycles as in Fig. 9.

(c) The Gerber line may then be drawn for a life of N_1 cycles as shown in Fig. 10 using $\bar{S}_{fe1}$ instead of Se.

(d) Using $\bar{S}_{fe1}$ in equation (18) and substituting r from equation (23) in (18) and rearranging gives

$$Sa_1^{\,2} + r^2 \frac{\bar{S}u^2}{\bar{S}_{fe1}} \cdot Sa_1 - r^2\, \bar{S}u^2 = 0 \quad (24)$$

which may be solved for the alternating strength Sa_1.

(e) The mean failure governing strength $\bar{S}f_1$ at N_1 cycles is then given by substituting Sa_1 in (19) to give

$$\bar{S}f = (Sa_1^{\,2} + Sm_1^{\,2})^{\frac{1}{2}} = Sa_1 \left(1 + \frac{1}{r^2}\right)^{\frac{1}{2}} \quad (25)$$

(f) The corresponding standard deviation σ_{Sf1} may then be obtained from experimental results of the type listed in (4).

(g) The loading stress values sa_1 and σ_{sa1} are synthesised into failure governing parameters along the stress ratio line of the diagram Fig. 10 to give

$$\bar{s}f_1 = sa_1 \left(1 + \frac{1}{r^2}\right)^{\frac{1}{2}} \quad (26)$$

and

$$\sigma_{sf1} = \sigma_{sa1} \left(1 + \frac{1}{r^2}\right)^{\frac{1}{2}} \quad (27)$$

(h) Provided that the failure governing stress and strength probability distributions along the stress ratio line are normal it can be shown that the safety margin (SM) is

$$(SM) = \frac{\bar{S}_{f1} - \bar{s}_{f1}}{(\sigma_{Sf1}{}^2 + \sigma_{sf1}{}^2)^{\frac{1}{2}}} = Z(f_1) \quad (28)$$

(to give SM > 0 for safety) consistent with equation (2).

(i) Inspection of Fig. 8 demonstrates that equation (28) implies the same safety factor (S.F.) on alternating stress, mean stress and failure governing stress, i.e.

$$\text{Safety Factor (SF)} = \frac{S_{f1}}{s_{f1}} = \frac{Sa_1}{sa_1} = \frac{Sm_1}{sm_1} \quad (29)$$

(j) It can be shown that the reliability or probability of survival for a life of N_1 cycles at this safety factor is then given by

$$R(N_1) = \int_{-Z(f_1)}^{\infty} \Phi \, dZ \quad (30)$$

(k) Equation (30) may be solved using tables for the area under a normal distribution.

An alternative method of estimating the reliability can be based on the probability distributions of cycles to failure on the S-N diagram shown in Fig. 9. The equivalent stress amplitude s_{fel} at zero mean stress for a load stress sa_1 and mean stress sm_1 can be derived from equation (18) which defines the Gerber line. For applied stresses this becomes

$$\left(\frac{sa_1}{s_{fel}}\right)^1 + \left(\frac{sm_1}{\bar{S}u}\right)^2 = 1 \quad (31)$$

Re-arranging equation (31) at $r = sa_1/sm_1$ gives the load stress amplitude s_{fel} at zero mean stress as shown on the S-N diagram shown in Fig. 10 where

$$s_{fel} = sa_1/\left(1 - \left(\frac{sa_1}{r.\bar{S}u}\right)^2\right) \quad (32)$$

The mean cycles to failure $\bar{N}_1$ at this stress level may then be read directly from the S-N diagram shown in Fig. 9.

At the stress level s_{fel}, the probability density function of the cycles to failure distribution, assumed to be normal for axial stresses is $f_1(N_1)$ as shown in Fig. 9.

The reliability or probability of survival for a life of N_1 cycles is then given by

$$R(N_1) = \int_{N_1}^{\infty} f_1(N)\, dN = \int_{Z_1}^{\infty} \Phi(Z)\, dZ \quad (33)$$

where $Z_1 = (N_1 - \bar{N}_1)/\sigma_{N1}$ (34)

and σ_{N1} = standard deviation of $f_1(N)$ (35)

Equation (33) may then be solved for the reliability, using tables for the area under a normal distribution.

4.3 Sequential Fatigue Loading of Tensile Member

It can now be shown that the reliability given by equation (33) may be used as a basis for estimating the reliability for cumulative fatigue loading conditions.

A tensile member subjected to an axial stress history involving a number of sequences of cumulative fatigue loads of the type shown in Fig. 11 is considered as an illustration. Kececioglu, Chester and Gardner (5) have examined alternating stresses applied sequentially assuming lognormal probability density functions of cycles to failure.

A method based on the procedure devised by Gardner (6) is now considered but the cycles to failure probability distributions for the axially loaded member are taken as normally distributed (4). The procedure suggested in (5) involves the assessment of an equivalent number of cycles at the second fatigue loading level representing the damage already incurred by the first loading and this is added onto the number of cycles at the second loading level. The same procedure can then be applied to any number of separate sequential cyclic loadings.

The reliability R_C of a tensile member for the first two loading cycles I and II of Fig. 11, is then given by

$$R_C = R(N_1 + N_2) = RI \times RII \quad (36)$$

where RI is the probability of survival of a member subjected to N_1 cycles at the first loading condition and RII is the probability of survival of a member subjected to N_2 cycles at the second loading condition and a damage equivalent of N_1 cycles at the second loading condition. This equivalent number is denoted as N_{1e}.

Using Bayes' conditional probability relationship, Gardner then showed that

$$R_C = RI \times RII = R_1(N_1) \times \frac{R_2\ (N_{1e} + N_2)}{R_1\ (N_1)} \quad (37)$$

or

$$R_C = R_2\ (N_{1e} + N_2) = \int_{(N_{1e}+N_2)}^{\infty} f_2(N)dN \quad (38)$$

where $f_2(N)$ is the probability density function of cycles to failure at the second stress loading level. By analogy with equations (33) and (34), the equivalent number of cycles at the second stress loading level to provide the same probability of survival as R_1 at the first stress loading level is obtained by solving the following equations for N_{1e} to give

$$R_1 = \int_{N_{1e}}^{\infty} f_2(N)\ dN = \int_{Z_{1e}}^{\infty} \Phi(Z)dZ \quad (39)$$

where

$$Z_{1e} = (N_{1e} - \bar{N}_2)/\sigma_{N2} \quad (40)$$

hence

$$N_{1e} = Z_{1e}\,\sigma_{N2} + \bar{N}_2 = Z_1\,\sigma_{N2} + \bar{N}_2 \quad (41)$$

since $Z_1 = Z_{1e}$.

Following this procedure, it has been shown in (7) that for 3 sequential loadings

$$Z_3 = [(N_{12e} + N_3) - \bar{N}_3]/\sigma_{N3} \quad (42)$$

where the cycles at the 3rd loading representing the damage due to the fluctuations of the first two sequences are given by

$$N_{12e} = Z_2\,\sigma_{N3} + \bar{N}_3 \quad (43)$$

and where

N_3 = cycles at third stress loading level

$\bar{N}_3$ = mean of density distribution of cycles to failure

σ_{N3} = standard deviation of $f_3(N)$.

If the coefficient of variation of cycles to failure distribution $f_i(N)$ is taken as

$$V_{Ni} = \sigma_{Ni}/\bar{N}i \quad (44)$$

it has also been shown in (7) that if the expression for N_{12e} equation (43) is inserted into equation (42), together with expressions for Z_2, Z_1 and N_{1e}, following the above procedure, then

$$Z_3 = \left[\frac{N_1}{\bar{N}_1}\frac{1}{V_{N1}} + \frac{N_2}{\bar{N}_2}\frac{1}{V_{N2}} + \frac{N_3}{\bar{N}_3}\frac{1}{V_{N3}} - \frac{1}{V_{N1}}\right] \quad (45)$$

and generally for K loading sequences

$$Z_K = \left[\left[\sum_{i=1}^{K}\frac{N_i}{\bar{N}_i}\frac{1}{V_{Ni}}\right]V_{N1} - 1\right]\frac{1}{V_{N1}} \quad (46)$$

If $V_{N1} = V_{N2} = V_{N3} = V_N = \sigma_N/\bar{N}$, then

$$Z_3 = \left[\frac{N_1}{\bar{N}_1} + \frac{N_2}{\bar{N}_2} + \frac{N_3}{\bar{N}_3} - 1\right]\frac{1}{V_N} \quad (47)$$

and generally for K loading sequences

$$Z_{i=K} = \left[\sum_{i=1}^{K}\frac{N_i}{\bar{N}_i} - 1\right]\frac{1}{V_N} \quad (48)$$

Using the procedures outlined in section 4.2, $Z_{i=K}$ derived in equations (46) or (48) may be used to determine the structural reliability for a specified life accounting for the order of loading with sequential axial loading conditions.

The summated term in equation (48) will be recognised as the cumulative usage factor for cumulative fatigue damage due to Palmgren-Miner (8)(9). Equation (48) then provides a direct correlation between the well known hypothesis usually known as Miner's Law which is used extensively for initial design estimates and the structural reliability dependent on the order of axial loading of a member at the same point in life. It also suggests that a more accurate cumulative usage factor, based on equation (46) might be given by

$$\left[\sum\frac{N_i}{\bar{N}_i}\frac{1}{V_{Ni}}\right]V_{Ni} \leqslant 1 \quad (49)$$

which should provide a statistically improved damage assessment than that provided by the customary Miner's Law expressed as

$$\sum\frac{N_i}{\bar{N}_i} \leqslant 1 \quad (50)$$

The methods presented in this section are restricted to indicating the basis for initial design estimates of the fatigue reliability for a specified life in cycles for axially loaded members and a comparison with existing deterministic design methods is provided. The general approach might be applied to initial design estimates for bolts, rivets, straps, studs or spot welds or generally axially loaded structural members with the appropriate empirical factors applied. A demountable joint has been used in (7) as an illustration of the approach.

Approaches are provided in (3) and (10) relating to design for the fatigue reliability life of components subjected to fluctuating combined bending and torsional loads as would be applicable for example to shaft drives.

5 CONCLUSIONS

For reasons of time and space we have confined ourselves to just two of the many possible failure mechanisms of mechanical items. Even so, the treatment has had to be general, and special features that apply to specific cases have had to be ignored. However, it has proved possible to show that design for reliability is possible with very little change to existing techniques. It has been shown that for both stress-rupture and fatigue failure mechanisms designers can continue to use their familiar factor of safety. Design for reliability means essentially that these have to be chosen on a rational basis which allows for variability in strength and load, as indicated in the body of the paper, rather than on a totally empirical basis. Even so, with the same factors involved, direct comparison of empirical and stochastic values is possible, and thus past empirical experience is not being thrown away. Furthermore it has been shown that in the case of fatigue the frequently used Miner's law can still be used with what is comparatively little modification (see equation (49)) for assessing cumulative damage with variable loading. So far as the authors can see there is no reason why these techniques should not be introduced into any design office immediately. That does not imply that we see them being used exclusively, but rather in parallel with existing methods, gradually taking over as experience is gained. With the accumulation of experience and confidence more advanced statistical techniques can be introduced.

To the designers we would say: customers are becoming more and more reliability conscious and future designs must acknowledge this. Deterministic design must give way to stochastic design. Some of the tools to achieve that goal are already available. To the reliability statistician we would say: new design techniques must be introduced to accommodate modern developments in stress analysis, and these should not be allowed to grow without also having a statistical foundation. Statistical methods must relate to design practice. It is crucial that these trends are co-ordinated. We believe that we have shown it to be possible.

REFERENCES

(1) CARTER, A.D.S. A view of mechanical reliability. Symp. Mechanical Reliability, IPC Science & Technology Press, 1980.

(2) COMMISSION OF THE EUROPEAN COMMUNITIES REPORT NO. COM(83), 350 Final, 1983.

(3) KECECIOGLU, D. Reliability analysis of mechanical components and systems. Nucl. Eng. Design, 19, (1972), pp. 259-290.

(4) KECECIOGLU, D., CHESTER, L.B. Combined axial stress fatigue reliability for AISI 4130 and 4340 steels. Paper No. 75-WA/DC-17, ASME Winter Annual Meeting, Houston, Texas, Nov/Dec, 1975.

(5) KECECIOGLU, D., CHESTER, L.B., GARDNER, O.E. Sequential cumulative fatigue reliability. Annals of the Reliability & Maintainability Symposium, 1974, pp. 533-539.

(6) GARDNER, E.D. Reliability of components subjected to cumulative fatigue. University of Arizona, Aerospace & Mechanical Engg. Dept., Master's Degree Report, 1971.

(7) KINKEAD, A.N. and MARTIN, P. An approach to the fatigue reliability of sequentially loaded demountable joints. Proceedings of the 8th Advances in Reliability Technology Symposium, NCSR and The University of Bradford, April 1984.

(8) PALMGREN, A. << Die lebensdauer von kugellagern >> ZYDI, Vol. 68, 1924, pp.339-341.

(9) MINER, M.A. Cumulative damage in fatigue. J.Appl.Mech., Vol. 12, Trans. ASME, Vol. 67, 1945, pp. A159-A164.

(10) REETHOF, G. Presentation during the discussion on the subject of research needs of reliability and failure prevention. Part of ASME Century 2 Conference, Reliability Stress Analysis and Failure Prevention Methods in Mechanical Design, San Francisco, 1980.

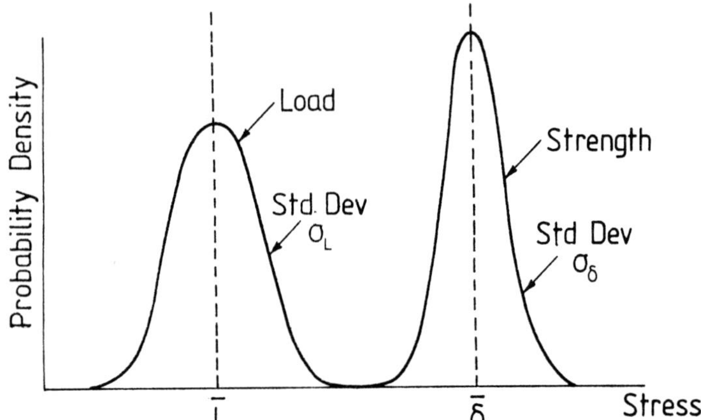

Fig 1 Relative location of load and strength distributions

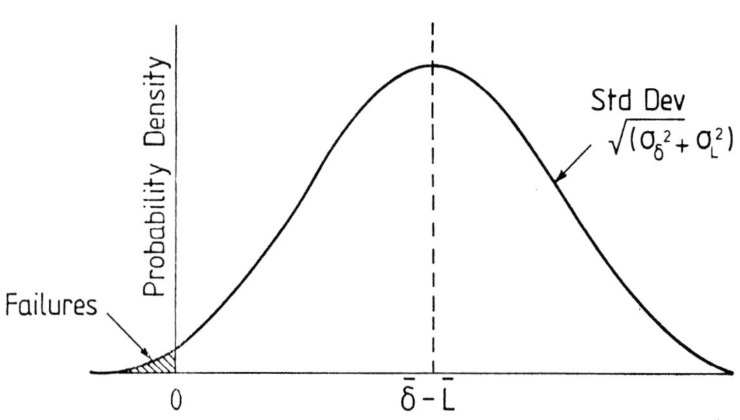

Fig 2 Load/strength difference distribution

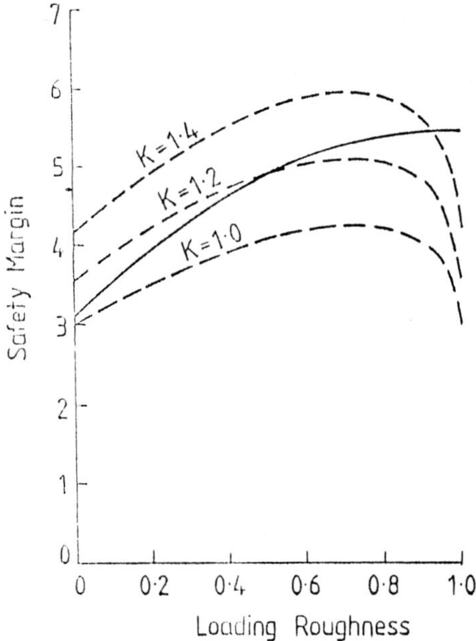

Fig 3 Comparison of design criteria

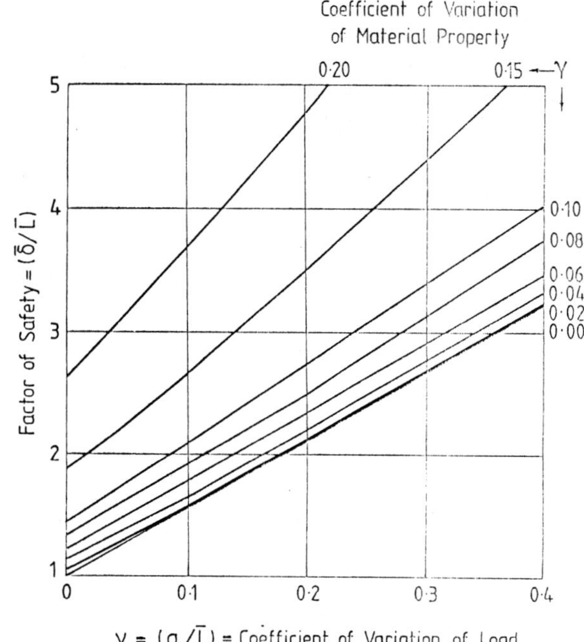

Fig 4 Minimum factors of safety to provide intrinsic reliability with near-normally distributed loads and strengths (Weibull-shaping parameter = 344)

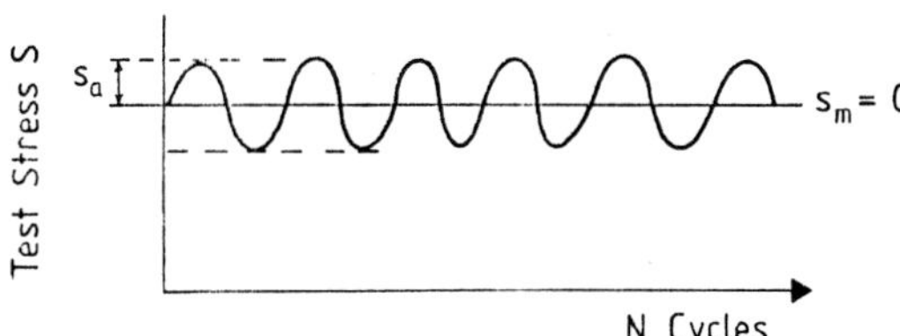

Fig 5 Cyclic fatigue test

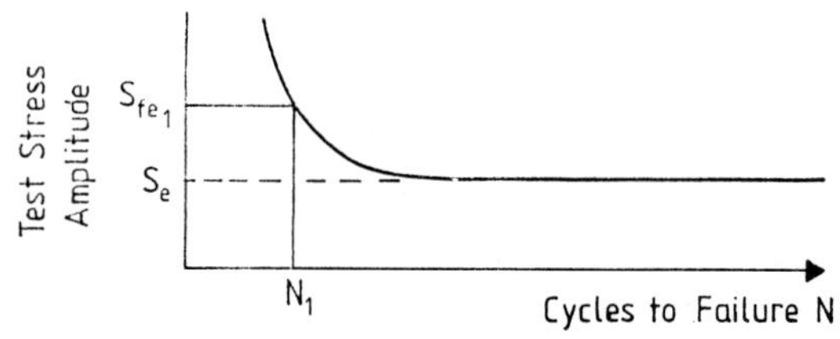

Fig 6 S–N test result

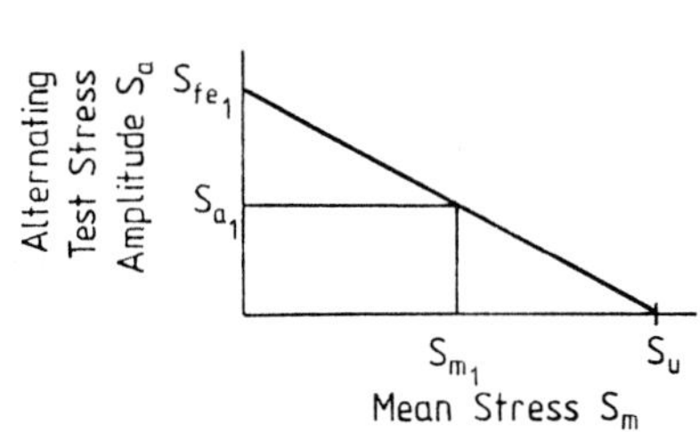

Fig 7 Modified Goodman diagram

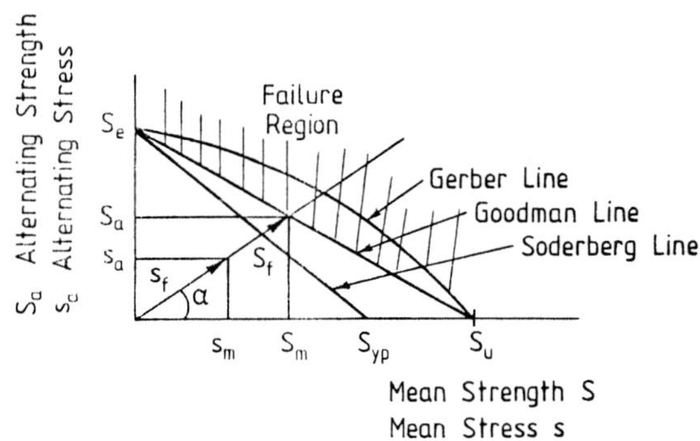

Fig 8 Design stress parameter

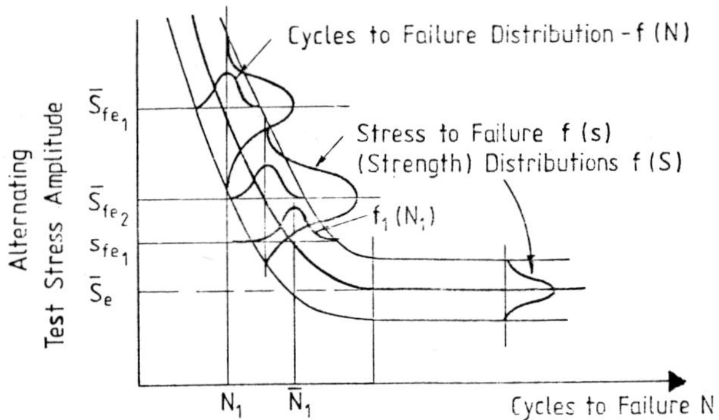

Fig 9 Distributed S–N test results

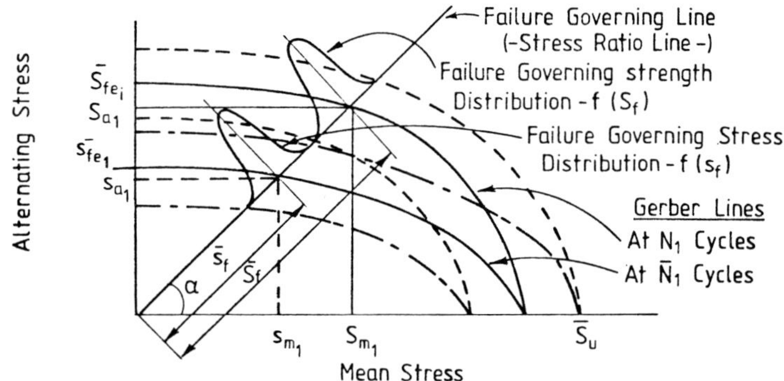

Fig 10 Distributed Gerber lines on modified Goodman diagram

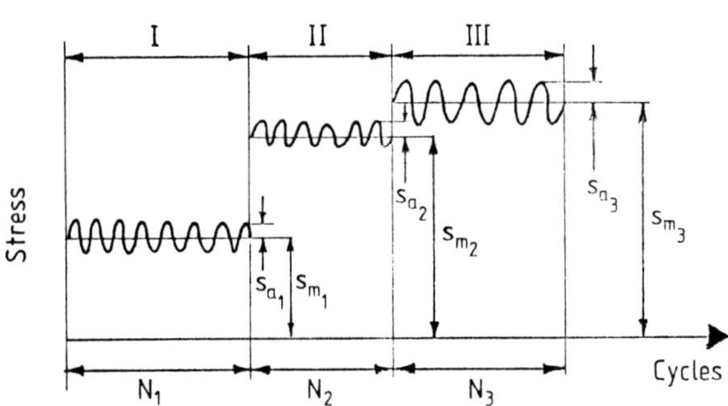

Fig 11 Stress history of component subjected to cumulative axial fatigue loads

Component based prediction for mechanical reliability

A O F VENTON, CEng, MIMechE, FIMarE, FSRS and T R MOSS, MPhil, CEng, FIMechE, FIQA, FSRS
RM Consultants Limited

SYNOPSIS Failure rate prediction for mechanical equipment is necessarily based on a smaller body of data than that for electronic equipment. The paper seeks to compare methods and results in the two fields to establish the extent to which methods used for electronic equipment may be applied in the mechanical case. It outlines two approaches to mechanical prediction using methods related to electronic parts count, which, though time-consuming have been shown to be effective. Both involve deeper analysis and give at least some consideration to failure modes. It also expounds in some detail a method used by the authors which takes stress into account in predicting failure mode rates for system components in specific applications.

The general conclusion is that failure rate prediction for mechanical equipment is possible, that it can achieve precision at least as good as electronic parts count methods and in the stress conscious mechanical case, probably better.

1 INTRODUCTION

It is often said that reliability in electronic engineering is more manageable than mechanical engineering. The grounds usually advanced are greater standardization of parts, larger numbers of parts and therefore a better statistical basis, and more predictable loading under normal operating conditions. It is implied that prediction of reliability in electronic equipment is easier and more trustworthy and that altogether mechanical engineering is a poor relation in statistically based prediction.

Moreover, it is argued that mechanical failures differ in nature from electronic failures because where the latter are invariably concerned with a chance excess of applied stress over inherent strength the former are often the result of the physical removal of material by wear. Thus electronic failures can be treated as random events in time while mechanical failures tend to be time dependent.

In reality chance overstress failures occur in mechanical parts as well as in electronic ones. Whilst wear, in the sense of the physical removal of material, is very unusual in electronic parts it is not unknown and one of the principle causes of mechanical failure, chemical change (corrosion), is a significant cause of electronic failure. Such causes as these are time dependent in both cases. In fact the failure theory presented by Haviland (1) attributes failure in both fields to time dependent deterioration stemming from growth of minute flaws, the physical imperfections inseparable from artefacts, in the material from which the parts are made.

2 PRECISION OF PREDICTIONS

Though there is clearly much truth in the contention that numbers of parts and standardization provide a much better basis for statistical prediction in electronic than mechanical equipment, the broad similarity of failure mechanisms suggest that similar methods should be applicable. The real test must lie in comparison of predicted failure rates with subsequent field experience in both categories.

Green and Bourne (2) report a comparison of predicted and observed values for reliability parameters for about 50 different electronic and instrument system elements. They expressed the comparison for each item as the ratio of the predicted value to the observed value. They found that the chance of that ratio lying within a factor 2 of the median of all the ratios was 70 per cent and within a factor 4 was 96 per cent.

Snaith (3) records an investigation of 130 diverse equipments and systems. They consisted of electronic, electro-pneumatic and mechanical equipment and control, instrument and mechanical systems. Table 1 shows his findings for the whole set of systems and equipments and for two equipment subsets. Also included for comparison are the findings of Green and Bourne. There is reasonable agreement between the two sources and Snaith's results show fair consistency between the whole set and the subsets. The extreme values of the factors, however, suggest that there may be a tendency towards a reduction in the spread of the comparison ratios as the set of items concerned becomes more homogeneous.

Snaith examined a number of the systems and subsystems in more depth. The results are shown in Table 2. Though the quantities concerned are not always failure rates, failure rates are major contributors to each one and the range of factors is meaningful in a broadly comparable sense. For the large chemical plant CO_2 shift contains a number of subsystems and the range of factors quoted arises because there was some doubt about which subsystem failures has been included in the actual system values. Similarly a range of values is given for instrumentation control valves because it was not clear to what extent failure of valves as opposed to control mechanisms had been included in the field data. In general, however, the range of factors for these mechanical systems is broadly similar to those obtained for electronic and electro-pneumatic equipments. The inference is that the causes of the difference between predicted and observed have not dissimilar weights in both cases.

3 PARTS COUNT PREDICTION

MIL - HBK - 217D (4) advocates a method of predicting electronic equipment failure rates based upon data given therein. It consists of establishing the failure rate for each generic part in the equipment, counting the number of each generic part, multiplying that number by the failure rate for each part and summing the resulting failure rates to give the estimated rate for the equipment. The process depends for its precision on the quality of the generic failure rates. These are derived from the model

$$\lambda_p = \lambda_b \, (\pi_E \, x \pi_Q)$$

Where:

λ_p = Predicted failure rate

λ_b = Base failure rate for the generic part

π_E = An environmental factor

π_Q = A quality factor

π_Q relates to sources and specified quality, π_E to the general environment in which the part will be employed. There are a substantial variety of the latter and they cover the range from ground benign through a number of ground sea and air applications up to the most severe, missile launch. The generic part values themselves are based on a large amount of data drawn from laboratory, development and field sources.

A companion source exists for mechanical generic data (5). It subdivides the data by source so that the environment is known, but does not classify by quality. With its aid however, a parts count for a mechanical equipment can be performed, taking into account the environmental effects, providing that all the required data are in the lists.

4 THE SPARSE DATA CASE

If data are few other means may be needed. A method of estimating the reliability of mechanical equipment when data are sparse has been advanced by Innes and Hammond (6). The process combines subjective weighting of part failures and failure modes with objective data for a very small number of the failures, or failure modes, to arrive at a prediction for failure, defined as loss of function, at the equipment level. To do this the equipment is analysed, using a fault tree, to establish the logical relationship between part failures or failure modes and loss of equipment function. Part of such a tree is shown in Fig 1. The complete tree is developed down to part or part failure mode level. At each gate subjective estimates are made of the relative probability of each input event, using service engineers with relevant experience to make individual estimates, and then combining their estimates into a single value for each event.

With a fully weighted tree it is possible to take one piece of hard data, which relates to one part failure or part failure mode and use the subjective weightings to propagate this upwards through the tree to arrive at a value for the equipment failure. Preferably several pieces of data are used in this way and the resulting estimates for the equipment combined statistically to provide the equipment failure rate estimate.

Innes and Hammond report in detail an estimate for a torque wrench performed in this way and record two more such estimates. Of the three two resulted in good agreement with field data.

5 PUMP FAILURE PREDICTIONS

One of the authors was responsible some time ago for a study utilizing Innes and Hammonds' method. It made estimates for five pumps designed by different manufacturers for the same duty. The values for three of them were similar and within a factor 2 of the best estimate which could be made from relevant field data about whole pumps. The remaining two values were significantly lower but similar to each other.

One example of each of the five pumps was tested on a severe version of the duty cycle, two of them being judged unsatisfactory and withdrawn from the test. There were the two which had yielded the low estimates. However, too much should not be read into this because only one ceased to function before withdrawal, so the connection between the study results and the withdrawal of the pumps from test was tenuous.

The overall conclusion from this study was that the method, though time-consuming could produce results as good as might have been expected from the use of parts count methods for an electronic equipment.

6 THE CRITICALITY ANALYSIS APPROACH

The US Department of Defense has set a standard (7) for the conduct of Failure Mode, Effect and Criticality Analyses at functional and hardware levels. At the hardware level, when considering an equipment, the Failure Mode and Effect part of the process consists of considering each part, postulating its failure modes and deducing their effects on the function of the equipment as a whole. This part of the analysis is entirely qualitative, with only the broadest classification of the severity of the effects. A formal tabular approach is used, though column headings are adapted to the needs of a particular analysis. An example of a typical set of column headings and set of responses is shown in Table 3.

The subsequent Criticality Analysis introduces quantitative measures for the criticality of individual effects, which are in reality estimates of the probability of the effect occurring during the specified operating time. An example of a set of column headings and a typical set of responses is shown in Table 4.

7 A GOVERNOR PREDICTION

A study of this kind, in which one of the authors was involved, concerned a governor for a steam turbo alternator. In this device speed and load were sensed and an error correction signal generated by an electronic unit; its output signal converted to hydraulic power by an electro-hydraulic servo unit; and hydraulically actuated steam control valves positioned by the hydraulic power. A failure mode, effect and criticality analysis was carried out to ensure that no intolerably hazardous condition could arise due to failures within these assemblies. Since criticality was involved, criticality numbers, i.e. a measure of the frequency of each individual effect, were calculated using data based mainly on Refs 8 and 9, and making due allowance for environment. The sum of those, for the effects which constituted malfunction amounting to failure are a measure of the failure rate for the system, though the production of that failure rate was not the real purpose of the study.

Parts count estimates based on Ref 8 had been made for two of the units before the F.M.E.C.A. began. Comparison between these estimates and the F.M.E.C.A. estimate of failure are shown in Table 5. The electrical component included in the electro-hydraulic unit rate is so small that it contributes less than one tenth of one per cent to the failure rate. Hence this unit can be regarded as a mechanical device.

Unfortunately we have no means of directly comparing the estimates with field data. The system has not yet operated in service. At best it can only be said that the results were credible.

An interesting feature though is the difference between the parts count rate and the criticality analysis rate. Parts count uses a generic failure rate for a part, modified by a factor to allow for the effect of environment. It may also be modified to take part-quality into account. Criticality analysis on the other hand takes the modified part failure rate, subdivides it into rates appropriate to individual failure modes and then uses the rates for the failure modes of interest only. Since, in a particular case some part failure modes may not be able to occur, or may have no effect, the assembly failure rate estimated therefrom will invariably be less than the parts count estimate. Also the failure mode set against which the failure rate is broken down very often contains a category 'other' which contains unspecified failure modes. Since only failure modes identified by the failure mode and effect analysis, for which specific values are found, are included in the prediction the 'other' category is excluded. There is however, the sneaking fear that there may be something contained within it which would be relevant, but since it cannot be identified, it cannot be used.

Overall, this aspect must lead to a degree of optimism in the results, however, the comparison suggests that parts count predictions are somewhat pessimistic because they will include inappropriate failure modes. As for the likely degree of pessimism, this case suggests that it may be similar for electronic and mechanical items; the former showing a 53 per cent increase of the criticality analysis rate and the latter 65 per cent.

8 PARTS STRESS PREDICTION

The environmental factors associated with generic data in parts count predictions are based upon observations of the failures which occur in the generic part in those environments. Thus they are taking into account the generality of operating stresses, internal equipment stress and externally imposed environmental stresses. At a stage in the equipment design when operating loads and internal environmental loads may be determined in more detail the generalization implicit in the parts count approach needs more detailed examination, particularly in critical areas.

In the electronic case this is accomplished by a more detailed estimation of the part failure rate which takes into account the electrical stress of the circuit as well as the local environmental stress imposed by temperature.

A base failure rate is first derived from a model of the general form

$$\lambda_b = A\left[\left(\frac{S}{N_s}\right)^H + 1\right] \cdot e^{B\left(\frac{T + 273}{N_T}\right)}$$

where:

λ_b = base failure rate

A = adjustment factor for each different type of part (e.g. capacitor)

S = ratio of operating to rated voltage

N_S = stress constant

e = base of natural logarithms

T = operating ambient temperature ^oC

N_T = temperature constant

B = shaping parameter

G and
H = acceleration constants.

The general form reveals non-linear relationships between the base failure rate and the electrical stress measured by the voltage ratio and the local environmental stress measured by the operating ambient temperature.

The base failure rate is then modified to yield the part failure rate using a simple product model which includes factors for environment and quality with others relating to the application stresses of the particular generic part concerned. As an example the factors relating to a transistor and a capacitor are shown in Table 6.

9 PARTS STRESS DATA

MIL-HBK-217D provides tabulated data for the base failure rate and all factors so prediction of part failure rate is relatively straightforward. The derivation of these tabulated data, in which the same base failure rate variable or factor can take on different values for each type of each generic component, implies a raw data base of astronomical proportions. The element of standardization, the enormous amount of laboratory test and development work undertaken with these parts and the field experience gained with them has made it possible.

Though Ref 8 provides mechanical data sufficient for an environmentally conscious parts count approach it does not present data or models which would allow part stress methods to be used. Neither does it break down part failure rates into failure mode rates though general information on that point can be found in, for example, Ref 9, which also includes some failure mode percentage figures and some models.

10 GROUPED PARTS AND SYSTEM COMPONENTS

One approach to what is currently available is to think in terms of groups of parts rather than individual parts. In fact in the F.M.E.C.A. outlined above it was found to be necessary to do this because so many mechanical failures result from interaction between two or more parts, e.g. increased resistance to motion, seizure. Following this further leads to thinking in system component or small equipment terms.

A method used by the authors for system components like valves has yielded good agreement with results experienced in practice. It employs field data from as wide a spectrum of applications as possible to define failure mode proportions and the likely range of failure rates experienced in practice. These data are then used to predict failure rates for each of the major failure modes associated with the specific operational and environmental conditions relating to the system being assessed.

11 FAILURE DISTRIBUTION

The failure data available for an equipment type is considered initially to establish the overall failure distribution. Where times to failure are known the cumulative hazard method for multipli-censored samples described by Nelson (10) is preferred. Otherwise, if mean failure rates only are available, the method developed by Moss (11) may be employed. The latter method involves fitting the best theoretical distribution by eye to the ordered sample failure rates using standard probability paper. This plotting method is a useful technique for identifying outliers and for defining the failure rate range for each equipment type.

12 STRESS FACTORS

Stress factors are estimated for each failure mode and applied to calculate the probability of the highest recorded failure rate applying for each specified application. This probability statistic is used to select the most likely overall failure rate from the distribution. The overall failure rate is then weighted by the modal proportions to derive a modal failure rate estimate.

13 EXAMPLE PREDICTION

As an example consider failure rate prediction for a twelve inch air operated globe type control valve, for use in the main oil export line of an offshore production platform, employing data from the S.R.S. data bank. The failure modes of interest are 'fail open' and 'spurious failure'. Stress ratios are estimated for the various operational and environmental conditions which apply to the modes. For the fail open mode these stresses are assumed to arise from the effects of pressure, temperature, size, corrosion and vibration; the spurious failures from vibration only. Hence stress ratios may be derived using simple linear relationships or engineering judgement. The ratios for this case are shown in Table 7.

These stress ratios are applied to the central probability value as dictated by the type of distribution.

Thus, $P_{Fo} = (1 - 0.5^{3.2}) = (1 - 0.11) = 89\%$

$P_{Sp} = (1 - 0.5^{2}) = (1 - 0.25) = 75\%.$

The failure distribution for air operated globe type valves derived from 20 S.R.S. data bank samples is shown in Fig 2. Overall failure rates associated with the above probabilities derived from this distribution are respectively 54 and 38 failures/10^6hours.

Table 8, taken from a recent U.S.N.R.C. study (12) indicates that the proportion of failures in each of the above modes is:

Fail open	10%
Spurious failures	5%

Hence estimates of failure rates for this specific application are:

$Fo = 54 \times 0.10 = 6f/10^6$ hr.

$F_{sp} = 38 \times 0.05 = 2f/10^6$ hr.

Note: Values are rounded up in all cases.

This method of predicting failure rates has been applied by the authors to a number of different situations and equipments and has shown close agreement with rates experienced in practice.

14 WHITHER NOW

Is it possible to suggest a standard method for estimating failure rates in mechanical equipment towards which we might work. The methods of MIL-STD-217D are attractive if for no other reason than that they are widely accepted for electronic equipment. However, those methods totally ignore failure modes. In the mechanical case failure modes are very important and in the example given above could only be ignored at the cost of gross inaccuracy. One approach which seems to yield good results and would therefore justify further development, and which might provide a basis for the deliberate collection of data is that shown in the example above. For that to be used, however, or indeed any failure mode conscious method for mechanical items, means that there is a real need to ensure that any data bank for mechanical items records failure modes as well as component or equipment failures. Perhaps this is the first essential step to take.

REFERENCES

(1) HAVILAND, R. P. Engineering Reliability and Long Life Design, Van Nostrand, 1964.

(2) GREEN, A. E. and BOURNE, A. J., Reliability Technology, Wiley, 1972.

(3) SNAITH, E. R. Can reliability predictions be validated? Second National Reliability Conference, March 1979, National Centre of Systems Reliability.

(4) MIL-HBK-217D, Reliability Prediction of Electronic Equipment, ANSI, 1982.

(5) RELIABILITY ANALYSIS CENTER, Non-electronics Parts Reliability Data, Rome Air Defence Center, Summer 1981.

(6) INNES and HAMMOND. Predicting Mechanical Design Reliability using Weighted Fault Trees, Failure Prevention and Reliability Conference, Chicago, Sept. 1977.

(7) MIL-STD-1629A. Procedures for Performing a Failure Mode, Effect and Criticality Analysis, November 1980.

(8) Guided Weapon System Reliability Prediction Manual, MOD (PE) Report No. DX/99/013-00. (Now superseded by DEF STD 00-41, Part 3).

(9) Applied Reliability Manual for Guided Weapons Systems, MOD (PE) Report No. GR/77/075/100.

(10) NELSON, N. Hazard Plotting for Incomplete Failure Data, Journal of Quality Assurance Vol. 1, No. 1, January 1969.

(11) MOSS, T. R. The Failure Characteristics of Process Control Equipment, M. Phil. Thesis, University of Bradford, 1981.

(12) USNRC, Data Summaries of Licence Event Reports of Valves at US Commercial Nuclear Power Plant, NUREG/CR 1363.

Table 1 Deviation of Individual Values of r from Median Value

	Number of Items	Median r	Extreme Value of Factor	Chance r < Factor 2 %	Chance r < Factor 4 %
Electronic and Computer Control Equipment	30		6	64	93
Electro-pneumatic protective Channel Trip initiators	23		3	60	100
All Equipments and Systems in Study	130	1.05	10	63	93
Electronic System Elements (Green and Bourne)	50	0.76		70	96

$$r = \frac{\text{predicted value}}{\text{observed value}}$$

Table 2 Mechanical Systems - Deviation of Predicted from Observed

Item	Quantity	Factor
Nuclear Power Station		
System without turbo-alternator	Unavailability	+ 1.33
Turbo-alternator (100-250MW)	"	- 2
System with 100-250 MW TA	"	- 1.47
System with 500 MW TA	"	+ 1.68
Large Chemical Plant		
Steam Boiler System	Downtime	+ 1.07
CO_2 Shift	"	- 2.43/-1.06
Gasifiers	"	- 1.18
Instrumentation Control Valves	"	- 3.75/-1.90
Boiler Feed Pump System		
Study 1	Failure Rate	- 1.28
Study 2	"	- 1.27
Study 3	"	- 1.93
High Pressure Die Casting Machines		
Tool System	Failure Rate	- 2.23
Tool Closing and Locking System	"	- 2
Metal Injection System	"	- 1.8
Operating Control System	"	- 3.78
Ancillary Systems	"	- 1.8

\- represents under estimate

\+ represents over estimate

Table 3 FMEA - Example of Column Headings and Responses

Column Heading	Response	Response	Response
Entry Code	2.2.1	2.2.2	2.2.3
Reference	100	100	100
Description	Needle roller bearing	1/1	1/1
Function	Support and friction reduction of motor gears	1/1	1/1
Failure Mode	Wear	Partially seized	Seized
Local Effect	Incorrect meshing of motor gears	Increased resistance to motor gear turning	Motor gear turning prevented
Effect on Module Output	Speed reduced	Speed reduced	No output
Effect on System Output	No significant effect	No significant effect	Erratic operation of pilot valve and hence govern possible trips
Severity Class	IV	IV	II
Remarks			Trip chance 50%

Severity Class

I	:	Catastrophic	- May cause death or system loss
II	:	Critical	- May cause severe injury, damage or mission loss
III	:	Marginal	- May cause minor injury, damage or mission degradation
IV	:	Minor	- No injury or damage risk, but unscheduled maintenance or repair.

Table 4 Criticality Analysis - Example of Column Headings and Responses

Column Heading	Response	Response	Response
Entry Code	2.2.1	2.2.2	2.2.3
Reference	100	100	100
Description	Needle Roller Bearing	Needle Roller Bearing	Needle Roller Bearing
Failure Mode	Wear	Partially seized	Seized
Severity Class	IV	IV	II
Failure Effect Probability	1.0	1.0	0.5
Failure Mode Ratio	0.73	0.15	0.05
Part failure rate ($f/10^6$ hrs)	1.2	1.2	1.2
Operating Time	1000 hrs	1000 hrs	1000 hrs
Failure Mode Criticality Number	8.76×10^{-4}	1.8×10^{-4}	6×10^{-5}
Item Criticality Number	1.06×10^{-3} (Class IV)		6×10^{-5} (Class II)

Table 5 Parts Count : FMECA Comparison

Unit	Parts Count $f/10^6$ hr	Criticality Analysis $f/10^6$ hr	$\frac{\text{Parts Count}}{\text{Criticality Analysis}}$
Electronic Governor	45.5	29.8	1.53
Electro-hydraulic Unit	50	30.3	1.65

Table 6 π Factor Application - Examples

Symbol	Factor	Transistor (1)	Capacitor (2)
π_E	Environmental Factor	*	*
π_A	Application Factor	*	
π_Q	Quality Factor	*	*
π_{S2}	Voltage Stress Factor	*	
π_c	Complexity Factor	*	
π_{SR}	Series Resistance Factor		*
E	Missile Launch ÷ Ground Benign	40	20

(1) Conventional Transistor Group 1

(2) Tantalum Electrolytic Capacitor

Table 7 Stress Ratios

Quantity	Basis	Ratio	Fail Open Mode	Spurious Failure Mode
Pressure	$\frac{\text{P Operating}}{\text{P Design}}$	$\frac{1000}{1500}$	0.7	No effect
Temperature	$\frac{\text{T Operating}}{\text{T Design}}$	$\frac{40}{50}$	0.8	No effect
Size	$\frac{\text{S Design}}{\text{S Median*}}$	$\frac{12''}{8''}$	1.5	No effect
Corrosion	Eng Judgement		2	No effect
Vibration	Eng Judgement		2	2
Stress ratio product (k)			3.2	2

* Median size of valves in the SRS sample

Table 8 NRC Data - Air Operated Control Valve Failure Modes

Failure Mode	Butterfly	Diaphragm	Gate	Globe	Total
Won't Start/Move	3	0	0	19	22
Won't Close	6	1	0	23	30
Won't Open	5	4	0	22	31
Leakage	65	21	12	154	252
Crack	3	1	0	4	8
Physical Distortion	2	7	0	4	13
Fracture/Break	0	1	0	1	2
Breach	1	1	0	1	3
Out of Limits	4	0	2	22	28
Out of Adjustment	1	0	0	1	2
Spurious Operation	1	4	0	7	12
False Response	1	0	1	6	8
Undefined	1	2	0	2	5
Total	93	42	15	265	416

Leakage	62%	
Physical Phenomena (Physical Distortion/Crack/etc)	6%	26%
(Won't Start/Move/etc)	20%	
Drift (Out of Limits/Adjustment)	7%	
Spurious Operation/False Response	5%	

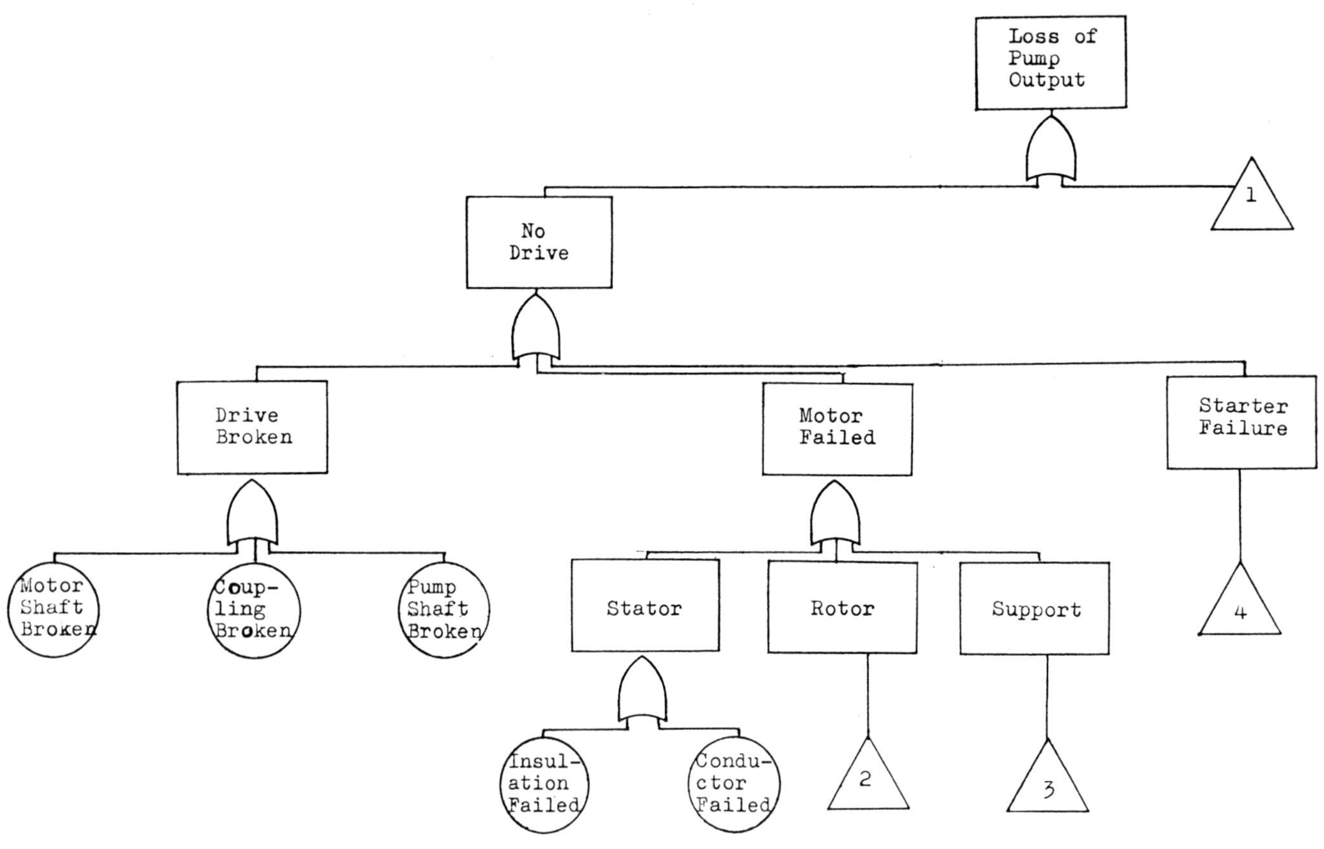

Fig 1 Pump fault tree – part

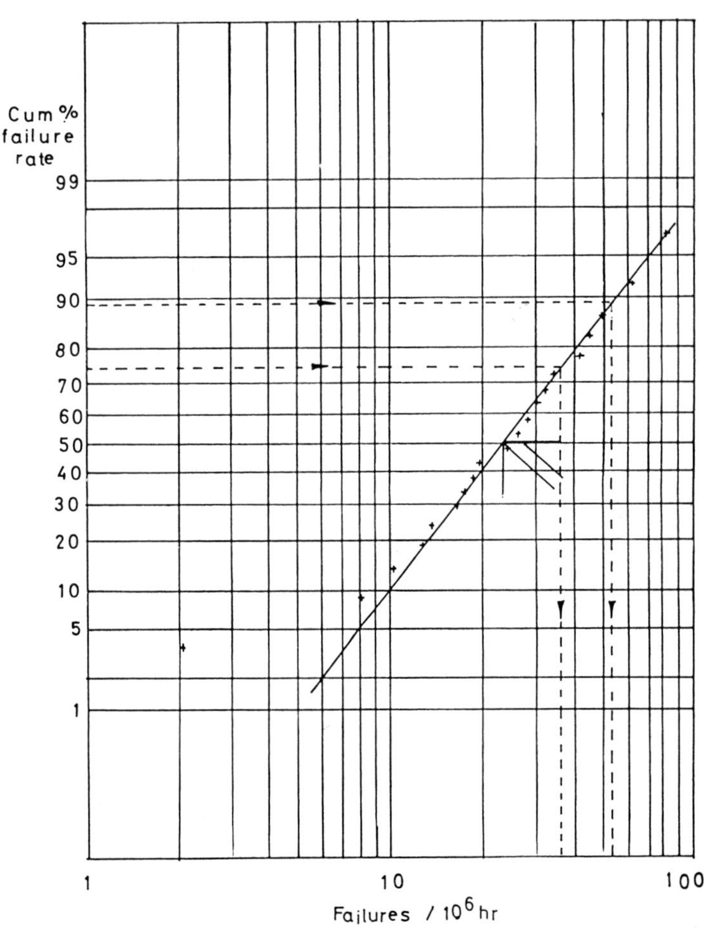

Fig 2 Air operated globe valves; distribution of failure rates

The influence of some tribological problems in mechanical reliability

R S SAYLES, M N WEBSTER and P B MACPHERSON *
Department of Mechanical Engineering, Imperial College of Science and Technology
* Also Westland Helicopters Limited, Somerset

SYNOPSIS

The role of tribology in mechanical reliability is discussed in general terms, and in the particular case of the U.S. nuclear power industry it is shown that tribological modes of failure have had a significant influence on around 25% of the major incidents experienced in recent years. The influence of tribology on the failure of particular components such as pump and valves are also investigated from published experience data and it is shown that tribological failure modes can be responsible for a relatively high proportion of component failures. This is shown to be as high as 80% for components such as water pumps.

Recent research results on tribological mechanisms and modes of failure are presented together with some new developments in condition monitoring. The significance of such research is discussed in the frame work of improving the overall mechanical reliability of components and systems.

1 INTRODUCTION

High reliability is essential where hazards are involved and we must be grateful to the industries involved with nuclear power, chemical plants and civil aviation for techniques and procedures which have allowed component system reliabilities to increase and grow as design and systems improve.

However, until quite recently reliability in relatively non hazardous areas of engineering has not been taken as seriously as we now know is required. Designs have evolved and then improved to get the best reliability from that system, rather than the basic design being evolved with reliability as a major design constraint.

With constantly reducing natural resources, the increasing costs of overdesign, a technique often used to try and achieve acceptable reliability, have imposed market forces which have meant many of Britains established engineering firms have gone out of business or been forced to impose radical changes in the philosophy of design and manufacture, and particularly to concentrate on quality and reliability within very competative world markets.

These trends have been evident throughout engineering in the UK, but not more so than in mechanical engineering where reliability considerations have often been glossed over in favour of more trivial and often purely asthetic considerations. Exceptions exist, and in some industries mechanical reliability has been recognised for many years as an important aspect of any product or system's market success. Rolls Royce cars represent a prime example of how quality and high reliability can be used to establish a strong position in world markets.

The reasons why mechanical reliability is so difficult to predict and assess, is now being generally accepted as due to the many and varied modes in which mechanical system can fail, and the many varied environments in which systems must operate. Because of the complexity of the problem reliability engineers have often been restricted to work with the purely statistical nature of mechanical failure and often been forced to ignore the root cause of reliability problems. Two approaches to improve mechanical reliability exist and are used; the first is to accept component reliability data and to employ redundancy and diversity to improve the overall system reliability. The second is more basic and has its roots in all aspects of engineering, and that is to improve basic component reliability. This latter approach is much the most difficult as it first of all involves a complete understanding of the mechanisms and physics of the failure modes involved. We need to study fracture mechanics, friction, wear, lubrication breakdown, etc, in order to influence the many modes of failure seen under the umbrella of mechanical reliability.

In practice reliability engineers employ an element of both these approaches; and although the latter is by far the most rewarding method of reliability improvement, the former is an excellent tool for identifying which components, and therefore to some extent, which modes of failure, are likely to lead to a maximised system improvement.

In this paper we set out to emphasise the potential gains available when our reliability resources are channeled into understanding mechanisms which we call tribological. It is our opinion that almost all mechanical failures can either be classified as a material failure or a tribological failure. We present evidence which shows the very high proportion of all mechanical failures which have a tribological origin. We also outline a small number of the many possible research programmes underway to understand and improve particular modes of tribological failure.

2 THE ROLE OF TRIBOLOGY IN INDUSTRY

The relative importance of tribology in industry depends largely upon the industry involved. Within the aerospace industry the premature failure of an important bearing or gear can be catastrophic and lead directly to complete system failure and very often loss of human life. As a consequence of the potential hazards of such failures, the aerospace industries invest heavily in research to understand and control tribological modes of failure. Some of the examples of our own research on specific failure modes set down later in this paper are in fact funded by the aerospace industries.

It is not easy to estimate the quantitative influence of tribological failures within such industries due to the lack of publicly available data of the detail required. However we would estimate that under all conditions of failure, and certainly where dangerous failures are concerned, the proportion of failures attributable to a tribological failure mode to be at least 40%, and more likely much higher than this, such that it would be fair to say that the majority of failures were tribological in nature.

With many industries, their interests in tribology are almost purely financial. Reduced friction and wear, longer bearing lives etc. simply mean improved efficiency. However the potential sums of money involved in such improvements are incredibly large.

The potential national gains available from improved reliability, and particularly through tribology, have been realised for many years and were brought to the attention of the government through the Jost report (1). This work led to the establishment of the UK Tribology Centres, set up to aid industry in reducing failure costs by improvements in tribological practices. A more recent I. Mech. E. report (2) has presented up to date financial information on the potential gains of improved Tribology and therefore improved mechanical component reliability. The potential financial gains are estimated to be as high as £700 M p.a.

In the chemical and nuclear power industries, tribological reliability has a dual influence, firstly through its effects on efficiency, and therefore profit, and also through its influence on safety. Because of the availability of data, particularly from the US nuclear power industry, it is possible to study and quantify the tribological contribution to overall reliability and risk. We have used this US data source to establish a quantitative estimate of the significance of tribological failures to nuclear plant, and also present the results as an approximate pointer to the relative importance of tribology within other similar industries.

2.1 Tribological failures in the nuclear power industry

It is important to identify and approximately quantify the influence of tribology on system and component failures within the nuclear industry. Data exists from the US and other sources to achieve this in a limited but useful way, and we have studied the data in two ways. Firstly, from the point of view of safety, by identifying events which were considered as significant faults in terms of their potential to lead to a core-melt situation, and secondly, by considering the general contribution tribology plays in the modes of failure of some of the more common mechanical components.

We are aware that tribology plays an important part in many other mechanical and non-mechanical components, but we do not at present have access to such data, if indeed it exists, to make any quantitative statement at this stage.

2.2 Tribological failures as significant initiating events in the U.S. Nuclear Power Industry

The United States Nuclear Regulations Commission (USNRC) require that any components or system failure or event occurring within the civil US Nuclear Power Industry be reported and documented. These Licensee Event Reports (LERs) are collated and analysed by the USNRC and act as an invaluable data base for component and failure analysis. We have used this data source to establish an estimate of the tribological influence on significant failures within US Nuclear BWR and PWR plants.

Oak Ridge National Laboratory (ORNL) were commissioned by the USNRC to study the potential severe core damage reflected by the events and failures seen within the LER experience data. A somewhat controversial status report by Minarick and Kukielka (3) was published in June 1982 and listed the events reported within the LER system between 1969 and 1979 in their apparent order of significance. From an initial 169 identified events or failures which had the potential to lead to a core-damage state, (these events/failures were called precursors in (3)). 52 were identified by ORNL to have an estimated probability of occurence of 10^{-3}/year or greater. ORNL classified these events as significant precursors, and they included well known events such as the recent Three Mile Island incident.

Response to the ORNL report, notably from the Institute of Nuclear Power Operations INPO (4) disputed many of the quantitative results (see (5) for a more detailed discussion), but in general, the ranking order of incidents was not questioned. We have used the ORNL ranking to estimate the influence of tribological failures on the mechanical reliability of US Nuclear Plants.

given by ORNL, 8 of the events are of a definite tribological nature involving lubrication system failure, seal failure, valve spindle siezure etc. Thus a lower bound estimate of the tribological influence can be set at around 8/52 or 15% of major incidents having a tribological initiating mode of failure. If we neglect the many purely electrical failures such as loss of offsite power through storm damage etc. and consider the tribological events within a population of significant mechanical and electromechanical failures. Then we have identified a reduced population of events. The tribological influence to major incidents now increases to 8/34, or 24%.

Many of the events reported by the LER system and listed by ORNL do not give enough information to identify the mode of failure of the incident's initiating event, and although these events are included in the total population, they have not been included in our estimates of the tribological influence. If we include a representative proportion of such undefined events (the proportional representation being based on our earlier estimates) then we have estimated that a fair reflection of the data would put tribological modes of failure as responsible for about 25% of all US nuclear power plant major incidents.

2.3 Tribological failures in some common nuclear power plant components

2.3.1 Valves

It is reported from the US nuclear industry (6) that about 30% of the total plant maintenance effort is devoted to valves. From data we have collected and analysed (6, 7, 8), the proportion of valve failures attributable to a tribological failure mode is around a minimum of about 31%, and most probably is higher than this, as many of the failure modes reported cannot be traced back to a basic mechanism of failure.

The U.S. nuclear industry data shows a minimum of 31.4% of the failure modes of valves to have a tribological origin. This calculation is based on a population of 1345 known failure mode, if we take account of 422 unknown causes of failure, and add to our figures a similar proportion as being probable tribological failures, the percentage increases to 41.3%.

Such statistics show the importance of tribological mechanisms, even to relatively passive mechanical components such as valves.

2.3.2 Pumps

With pumps the modes of failure are dominated by tribological mechanisms. A typical distribution of failure modes is shown in Figure 1. The data is taken from Dorey and Gachot (9) and represents the failure modes seen in the French nuclear industry from a population of 536 water pumps which had suffered 465 failures during a total operating time of 4.5×10^6 hrs.

Considering just the failures in bearings, seals, and lubrication systems, as the only tribological modes, we can see that these amount to a total of 64.5% of all possible failures. Considering 'other leakages' to be all caused by tribological modes of failure, then the proportion of all pump failures attributable to a tribological mode of failure increases to be in excess of 80%.

Similar figures can be derived from the chemical plant data analysis reported by de la Mare (10). This data was also derived from a relatively large population (669 repairs on a population of 79) and serves to emphasise the importance of tribological failure modes in pump performance.

3 RESEARCH ON SOME IMPORTANT TRIBOLOGICAL MODES OF FAILURE

The nature of tribological modes of failure are such that they can very rarely be designed out of the system in the way that we can often deal with other types of mechanical failure. The early Comet windows, Liberty ships, box section bridges, etc, are examples of mechanical material failures where an understanding of the mode of failure led to design changes which effectively removed the problem. Tribological failures are not like this, we often understand the mode of failure, but do not have sufficient control to remove the problem, the best we can often achieve is to extend the useful life, and much research is in progress trying to do this.

An alternative to controlling the failure mechanism, but not as satisfactory a solution, lies with the use of condition monitoring techniques. With many types of critical tribological component the onset of failure can often be detected by monitoring and analysing vibration patterns, or lubricant conditions. Techniques have been developed to achieve this and one particularly promising method employing on-line ultrasound analysis of lubricant and hydraulic systems has recently been reported by Harries et al. (11) and will be described in some detail.

3.1 Early life failure in rolling contacts

The classical analysis of rolling contact fatigue has historically centred around the assumption that the pressure distribution, in the contact zone, can be approximated by the semi-elliptical distribution obtained by Hertz (12). There have been numerous fatigue models presented, Popinceanu et al. (13) compare a number of these with experimental data. In practice it is found that under service conditions, bearings often exhibit several different failure modes, and failure times differ greatly from those predicted by classical analysis. Of particular importance are the so called early life failures, where a number of bearings fail long before the times expected from accepted failure distributions.

The presence of bi-modal effects seen in many of the published Weibul distributions for rolling contact fatigue, implies that there are at least two competing fatigue mechanisms. Sayles and Macpherson (14) and Harries et al. (15) suggested that the mechanism responsible for the early life failures observed, was surface, as opposed to sub-surface, crack initiation. Sayles and Macpherson (14) contamination present in the lubricating oil and an important conclusion from their work was that it is not the continued presence of debris in the lubricant, but the initial damage to the bearing surface, that leads to premature spalling failure. Figure 2 shows a three dimensional surface plot of a bearing surface taken from one of the test specimens used in (14). The surface damage from the 'rolled in' debris can clearly be seen. Evidence suggests that such defects are responsible for many of the early failures in rolling contacts and much research is in hand to understand the mechanisms involved.

An important element of the problem is understanding the influence of such surface defects on the pressure distribution and film thickness within the elastohydrodynamic (EHD) contact region. Cheng and Bali (16) have solved the governing coupled EHD equation, for line contact, with a mathematically idealised furrow or asperity, and Wedeven (17) has used an optical EHD test machine to measure departures from the classical EHD conditions, caused by a debris initiated dent, but to date, no complete solution of the problem has been put forward.

The authors have devised a numerical technique which allows the calculation of the normal contact pressures for a roller loaded against any surface profile. The program solves the general 2-dimensional contact problem, but differs from previous work as it incorporates a specialised algorithm to cope with the fact that the apparent contact area consists of many interelated contact spots. A full description of the technique is intended for a later publication, but some preliminary results are presented here.

The input for the program may be any digitised surface profile data; a digitised form of the analogue signal produced from a stylus surface measuring device for example. This may then be used as one of the bodies in the contact program. The choice of the roller dimensions, roller position, elastic constants and the total normal force are input parameters to the program. The system thus allows the pressure distribution to be calculated from the real profile data within a short space of time. The authors believe that this technique of loading real surface profiles in an elastic manner is unique.

3.2 Analysis of the pressure distribution around a real debris dent

The computer program described above has been used to calculate the pressure distribution around a debris dent produced during the experimental tests described in (14). The experimental values for the load, Young's modulus and roller dimensions reported in (14) were also used in order to reproduce, as nearly as possible, the true experimental conditions. Figures 3a-3d show the calculated pressure distributions superimposed over the smooth case theoretical results. The diagrams represent a time history of the normal pressures as the debris dent moves into and through the contact zone.

It can be seen from these results that pressures in excess of 2.5 times the theoretical maximum pressure can exist on the contacting surfaces. The analysis of the state of stress beneath the surface is at present being carried out; it is hoped to include the residual elastic stresses caused by the initial debris dent deformation in these stress calculations. However it seems clear that the assumption of a state of Hertzian stress being present in the contact zone is inadequate if a true picture of the fatique mechanism is to be derived. Furthermore the sharp peaks seen in the pressure distribution suggest that there will be significantly high shear stresses close to the bearing surface. This result would confirm the reported findings (14,15,17) that debris initiated fatigue is surface, and not sub-surface orientated, as in the classical analysis.

The effects of the presence of an EHD or partial EHD lubricant film, and the extent of local plastic deformation are the subjects of continuing research with this technique.

3.3 Ultrasonic Condition Monitoring

Recent work on filtration of oil supplies to rolling contact bearings (14) has shown that the presence of certain sizes of contamination particle and wear debris can lead to much reduced bearing fatigue lives. There is also evidence to show that the catastrophic failure of gear and bearing surfaces is often preceded by a slow surface degeneration process. This would go unnoticed but for an increase in wear debris within the lubricant system, and a possible increase in bearing noise. At present the most commonly used technique for monitoring the failure of components in this way is vibration monitoring, but this is often unreliable, providing inconvenient false alarms. As a more general system monitor, oil analysis techniques are employed such as ferrography. These techniques are slow and require lubricant samples being taken for laboratory testing. So, an on-line system that monitors lubricant quality, and particularly debris particle size, concentration, and ideally constitution could not only provide a superior method of identifying impending failure, but could also help prevent such failures.

An ultrasonic contamination monitor has been developed by Harries *et al.* (11) and relies on two piezoelectric transducers arranged in the geometry shown in Figure 4. One transducer acts as an ultrasound transmitter and the other, matched, transducer as a receiver. At MHz frequencies ultrasound is highly directional with a 5mm disc resonating at 5MHz having a divergence of typically 4° in liquids. The two intersecting cones of directionality form an effective measuring volume as shaded in Figure 4. Any acoustic discontinuity within this volume will scatter the transmitted ultrasound in all directions, some of which will be towards the receiver. The acoustic pressure P_s incident on the receiver is converted by the transducer to a voltage, proportional to P_s. This voltage will alternate at the ultrasonic frequency, with demodulation yielding an envelope voltage V_s. The received ultrasonic therefore proportional to V_s^2. The technique described by Harries et al. uses V_s to infer information about the contents of the measuring volume.

The ultrasonic transducer system is mounted in a closed loop flow rig (Figure 5) where the injection of a known concentration of fine particles can be used in conjunction with a HIAC optical particle counter to callibrate the system. Preliminary experimental testing of this technique has used glass microspheres in the size range 35-75μm dispersed in a low viscosity oil. The concentrations studied to date ranged from detection of single particles to 50 ppm. The results are very encouraging (Figure 6) and in good agreement with theoretical predictions. Tests with smaller particles, irregularly shaped particles, higher concentrations, and mixed flows of hard and soft (air bubbles) particles are currently in progress.

There are several practical problems to overcome before this prototype monitoring device can be used as a standard piece of on-line equipment, however the ultrasonic contamination monitor seemingly offers an ideal method for detecting the presence of wear debris in lubricant and hydraulic systems. It has the potential to do this on-line, at low cost, and with minimal alteration to existing mechanical equipment.

3.4 Micropitting in Gears

A particular failure mechanism, often found in gears, is initially seen by the development of a large number of small surface cracks, up to about 10μm deep, and running transverse to the rolling and sliding directions. The cracks have a further peculiar characteristic, in that they are inclined to the surface in a direction which changes on each side of the pitch line (Figure 7). On subsequent loadings of these cracks the edges break away and create a large number of small craters in the surface, the condition known as micropitting (Figure 8). This process can very often carry on indefinitely and leads to extremely high wear-rates or to destructive surface pitting fatigue. Running becomes increasingly noisy and imposes high stresses on other components (the situation cannot be tolerated for more than a few hours).

Recent research by Olver (18) at Imperial College has employed slip line field theory to show that following plastic asperity contact and local deformation, in the presence of surface sliding and therefore frictional traction, residual tensile stresses are created just under the surface which change direction on each side of the pitch line, i.e. corresponding to when the traction forces change direction. The residual stresses are predicted as being approximately normal to the direction of the observed cracks, and therefore on unloading of the contact these residual stresses would open up cracks in a preferential direction, and in such a direction that is in good agreement with the microcracks observed on gear tooth sections.

With the understanding of the micropitting mechanism presented by Olver it should be possible to define contact conditions, and particularly asperity geometry, which will minimise the effects of this particular tribological failure mechanism. Research is continuing on this problem and its influence on gear wear and reliability.

3.5 Relevance to the Process Industry

The relevance of tribological modes of failure in mechanical equipment from the nuclear and aerospace industries is outlined in this paper, and we have no reason to suppose the process industry is less susceptable to such problems. In order to remove such problems, or improve component reliability to acceptable levels, it is essential that we understand the engineering and physics involved in creating failures. We need to understand a problem before we can solve it. To achieve this, much research is needed to build a sufficient level of control, in order that future components and system designs can be improved.

In Section 3 of our paper we have outlined just a few of the many possible areas of tribological research where improved methods of system monitoring, and an improved understanding of failure mechanisms, is being established in order that improvements in mechanical component, and eventually, system reliability can be achieved. We present our work in this way in the hope that engineers and scientists within the process industry, and who are better equiped than ourselves, might find our results useful in improving reliability in their own areas of expertise.

4 CONCLUSIONS

Evidence is presented to define, and in some special but important cases, to quantify the degree of inflence that failure modes of a tribological type have on general mechanical reliability. The influence of tribological modes of failure on nuclear plant components is shown to be very significant, and in the case of water pumps as many as 80% of all failure modes experienced are the result of a failed tribological function.

The influence of these and other tribological failures are also examined in terms of their contribution to the significant potentialy dangerous events seen within the US nuclear power industry. We estimate that tribological failure mechanisms contribute around 25% of the initiating events which are seen as significant in terms of their potential to lead to a core-melt situation.

Results are presented on some of the research and development projects with which the author's laboratories are involved, and which are designed to enhance our knowledge of tribological failure mechanisms and improve component and system lives within the mechanical reliability field.

Early life failure in rolling contacts, such as those within ball and roller bearings are surface contact. Localised high contact stresses are shown to be created by the passage of a rolling element across a local surface defect. Results are presented which show the potential of a recently devised elastic contact analysis method to quantify these localised stress effects. The results shown represent a preliminary series of tests of a technique which has an exciting potential for investigating further surface contact, and therefore tribological, problems.

Gear tooth failure is often the result of a surface degeneration process known as micropitting. An explanation of this failure mechanism has recently been presented by Olver (18) and is seen as the result of residual subsurface stresses caused by local surface asperity plastic deformation. Such understanding will undoubtably lead to improved reliabilities within geared systems sensitive to this mode of failure.

Work is also presented which outlines a recently developed ultrasonic technique of lubricant condition monitoring (11). The ultrasonic method is shown to operate in lubricants and hydraulic systems, and to have the potential to monitor particle contamination continuously, and on-line. The system is shown to have the potential of helping to prevent some of the catastrophic tribological failures experienced within recent years, particularly in the aerospace industry.

5 ACKNOWLEDGEMENTS

The Authors would like to thank C.J. Harries and A.V. Olver whose works we have described in some detail, and also gratefully to acknowledge the financial support of RAE Farnborough, the U.S. Navy's Office of Navel Research, and Westland Helicopters in the research described.

REFERENCES

(1) Lubrication (tribology) education and research, (Jost Report), Department of Education and Science, HMSO, London, 1966.

(2) JOST, H.P. and SCHOFIELD, J. Energy saving through tribology: A techno-economic study. Proc. Instn Mech. Engrs , 1981, 195, pp151-173.

(3) MINARICK, J.W. and KUKIELKA, C.A. Precursors to potential severe core damage accidents: 1969-1979, A status report, ORNL, NUREG/CR-2497, ORNL/NSIC-182, 1982.

(4) ROSEN, S. INPO response to ORNL ref.(8) report, reported in Nucleonics Week, Oct 21, 1982.

(5) SAYLES, R.S. Assessment of the PSAs made for Sizewell 'B', Proof of evidence presented to the Sizewell 'B' Public inquiry, April 1982, Suffolk County and Coastal Distrct Council LPA/P/S.

(6) SPRUNG, J.L. (ed) Proc. EPRI/DOE workshop Nuclear Industry Valve Problems, printed by Sandia National Laboratories (SAND80-1887, UC-78), Jan 1981.

(7) HUBBLE, W.H. and MILLER, C.F. Licensee event reports of valves at U.S. Commercial nuclear power plants (Jan 1976 to Dec 1978) NUREG/CR-1363 1980.

(8) REYER, R.J. and RIDDINGTON, J.W. Valve failure problems in LWR power plants, 1981, Nuclear Safety, V22, No.2, pp229-235.

(9) DOREY, J. and GACHOT, B. Pump reliability data derived from electricite de France operating experience, 1978, ASME PVP-PB-032.

(10) DE LA MARE, R.F. Economics of breakdown of mechanical equipment, 1978, ASME PVP-PB-032.

(11) HARRIES, C.J., SAYLES, R.S. and MACPHERSON, P.B. An on-line ultrasonic device for monitoring contamination in fluids, 1984, to be presented at the Swansea conf. of Condition Monitoring, April 1984.

(12) HERTZ, H. On the contact of rigid elastic solids, and on hardness, miscellaneous papers, 1896, Macmillan, London.

(13) POPINCEANU, N.G., DIACONESCU, E. and CRETU, S. Critical stresses in rolling contact fatigue, Wear, 1981, 71, pp265-282.

(14) SAYLES, R.S. and MACPHERSON Influence of wear debris on rolling debris fatigue. Rolling contact fatigue testing of bearing steels, 1982, ASTM STP 771, ed. Hoo, J.J.C. pp. 255-274.

(15) HARRIES, C.J., WEBSTER, M.N., SAYLES, R.S. and MACPHERSON, P.B. Bi-modal failure mechanisms in tribological components, Reliability Eng'g, 1983, 4, pp169-180.

(16) CHENG, H.S. and MRINAL BALI Stress distributions around the furrows and asperities in EHL line contacts, Solid Contact and Lubrication Seminar, ASME, Chicago Nov. 1980, AMD-Vol 39.

(17) WEDEVEN, L.D. Influence of debris dent on EHD lubrication, Trans. ASLE, 1977, 21, 1, pp41-52.

(18) OLVER, A.V. Micropitting and asperity deformation, 9th Leeds-Lyon symposium, Lyon, September 1983.

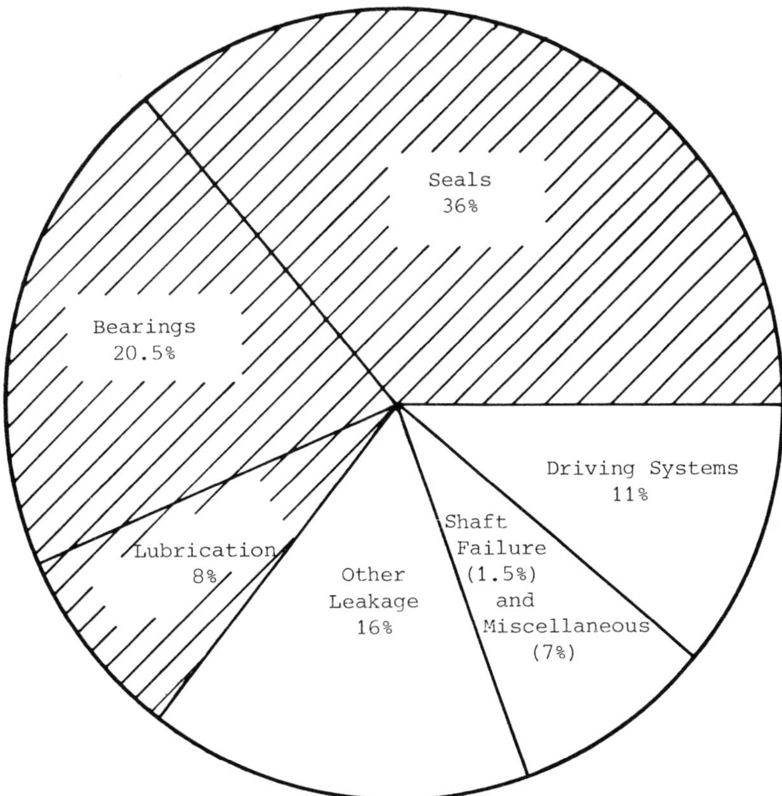

Fig 1 The distribution of failure modes seen in French nuclear power plant water pumps (Dorey and Gachot, ref. 8)

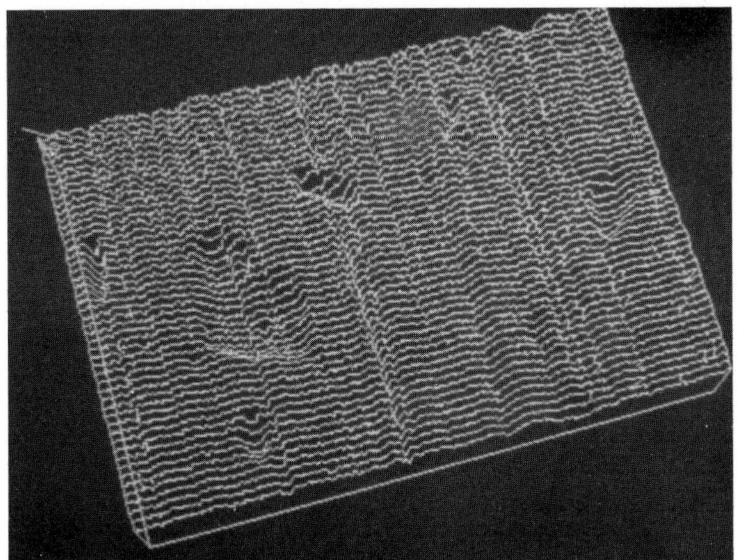

Fig 2 Talysurf parallel surface profile measurements representing 2x1.4mm of a roller bearing surface and showing the indentation damage caused by rolled-in wear debris. The main crater is about 2 μm deep, and the original machining marks can be seen running up and down the surface

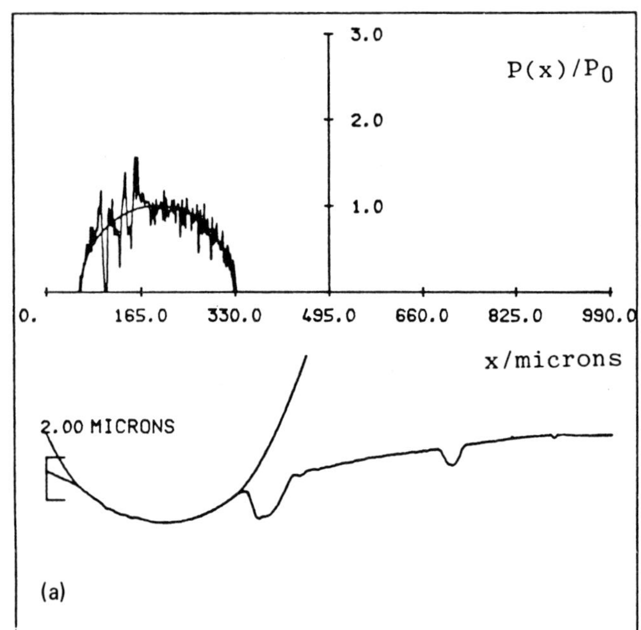

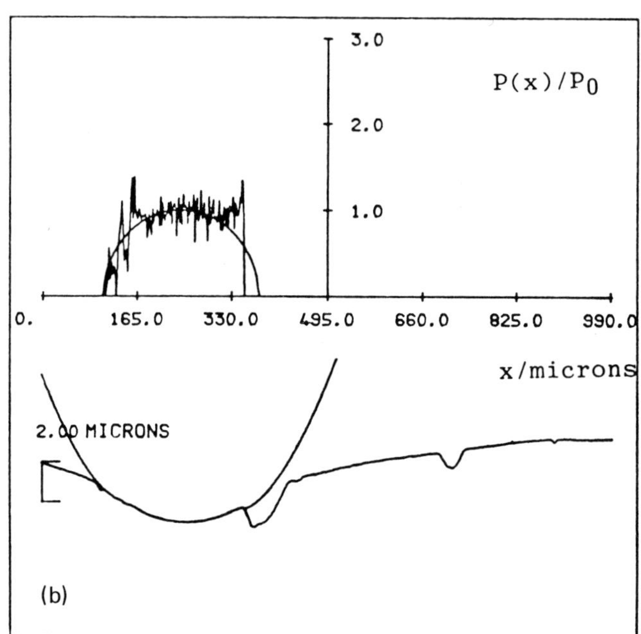

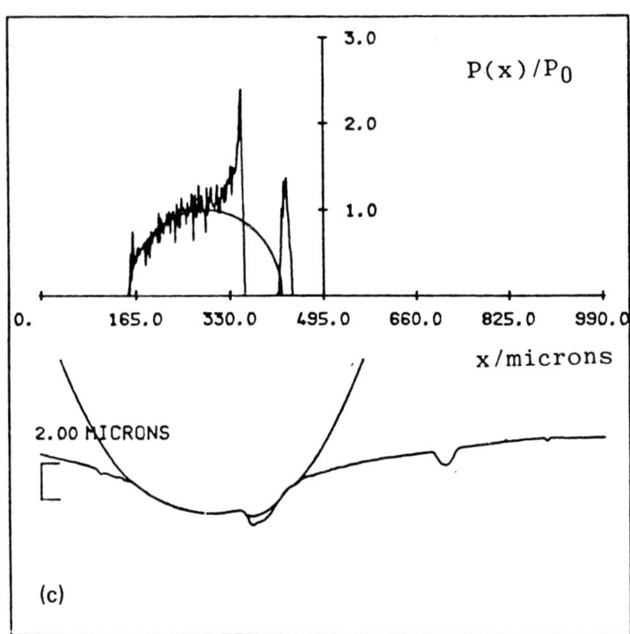

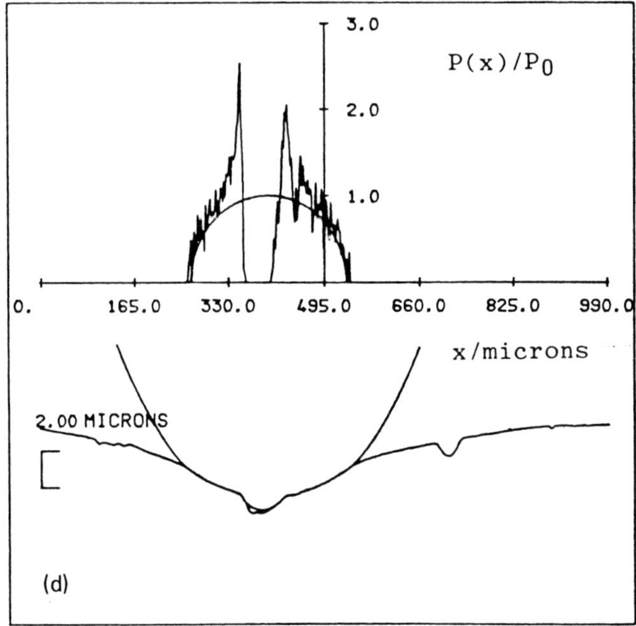

Fig 3 (a–d) Sections through the elastic contact geometry of a smooth roller loaded against the surface shown in Fig 2 as it moves across the debris-created dents. The Figs also show the normal elastic stress distribution compared to the smooth surface Hertzian theoretical distribution. Note the variations in stress created by the surface roughness and particularly the very high concentrations of stress as the roller experiences the relatively sharp edges of the debris dent

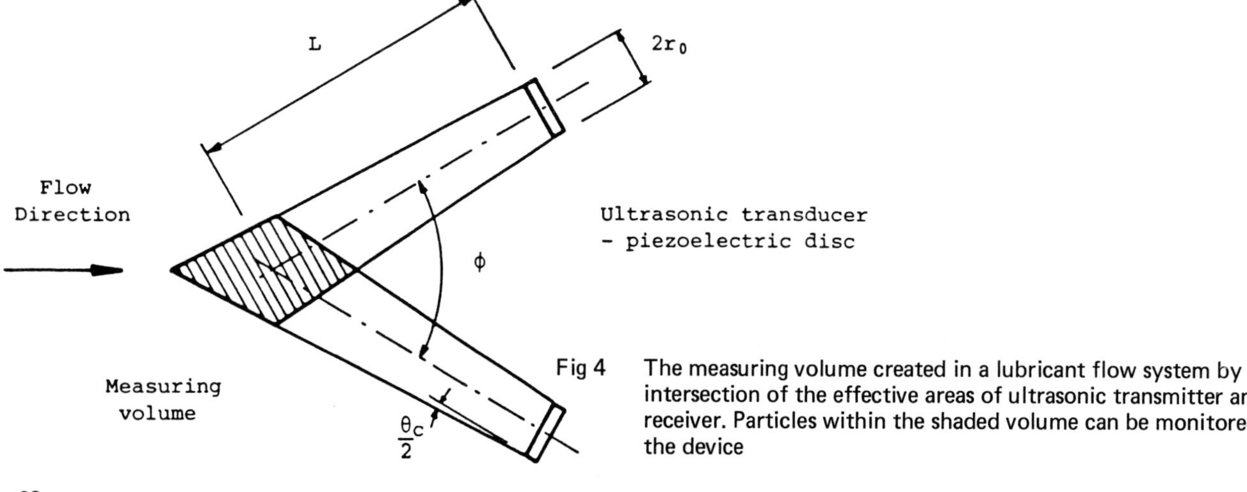

Fig 4 The measuring volume created in a lubricant flow system by the intersection of the effective areas of ultrasonic transmitter and receiver. Particles within the shaded volume can be monitored by the device

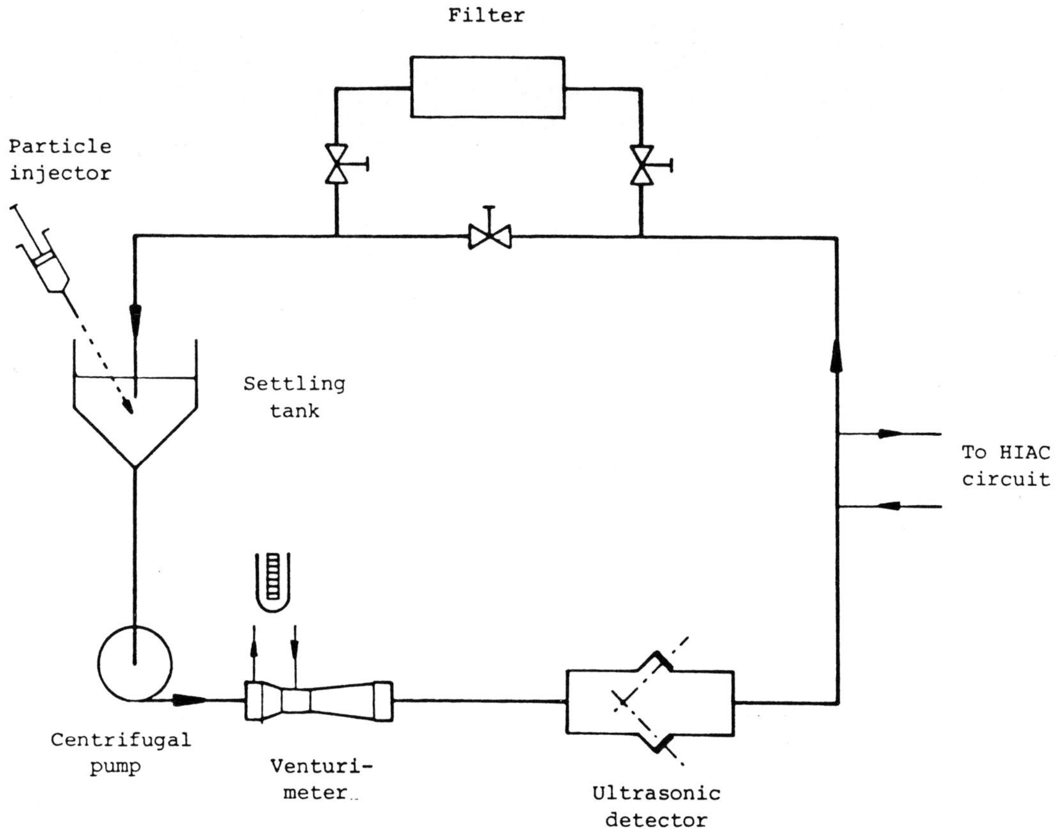

Filter
Particle injector
Settling tank
To HIAC circuit
Centrifugal pump
Venturi-meter
Ultrasonic detector

Fig 5 Schematic diagram of the closed loop lubricant flow rig used to investigate the ultrasonic condition monitoring system

Size distribution
+ = A
△ = B
◇ = C
▽ = D
$\bar{V}_s^2$
2.0
1.5
1.0
0.5
0
0
10
20
30
40
50
60
concentration (ppm)

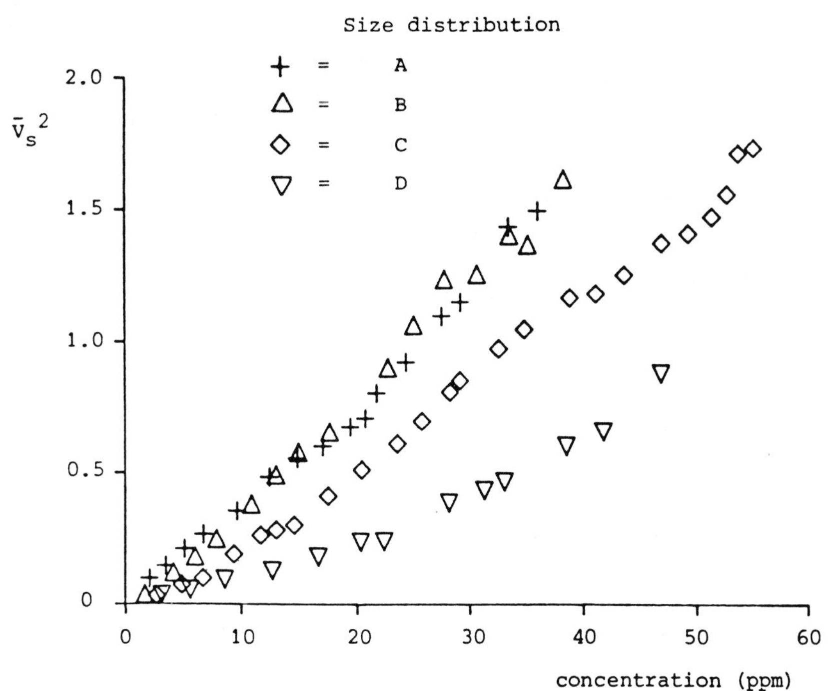

Fig 6 Preliminary results of correlations of the ultrasonic condition monitoring device. The ordinate is in units of $(\text{volts})^2$ and represents a signal property which has a theoretically linear relationship with the abscissa representing volume concentration. The experimental points represent particle size distributions (A-63 to 75μm, B-53 to 63μm, C-45 to 53μm, D-37 to 45μm). (The results are taken from Harries *et al.*, Ref. 14)

Fig 7 A section through the microcracks developed on a gear tooth. The cracks are about 10μm deep and change direction when the surface sliding changes on each side of the pitch line. (The scale of such sections prevents the portrayal of cracks on both sides of the pitch line in the same Fig) (x 164)

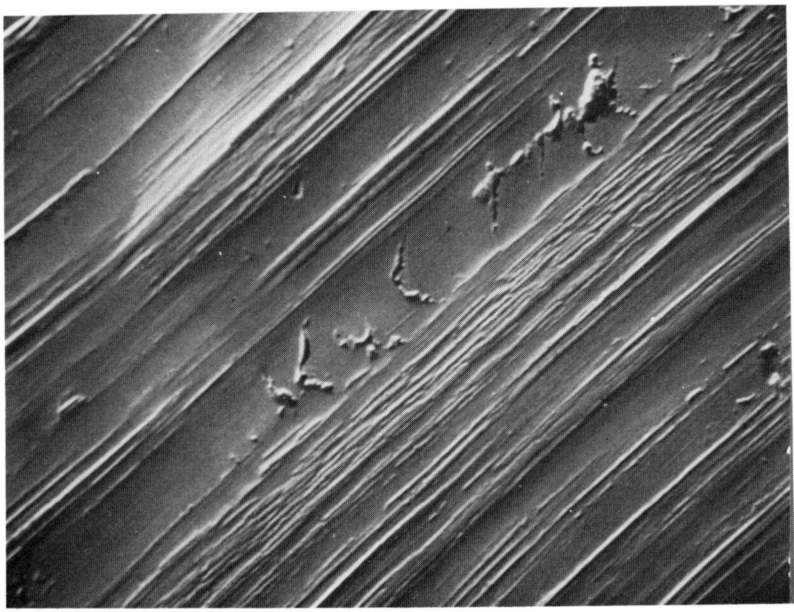

Fig 8 Electron microscrope view of a region just beginning to form micropits from the generation of small cracks on a gear tooth. The pit in the lower left hand quadrant of the Fig is about 15μm across (x 328)

Mechanical valve reliability

R J AIRD, BSc, CEng, MIMechE
T R MOSS, MPhil, CEng, FIMechE, FIQA

SYNOPSIS Valves play an important role in many process plants and valve reliability can be vital to the safety and availability of the plant. Many instances have been reported, in Europe and the United States, where valve failure has had serious consequences and this has resulted in major test programmes being established in several countries. The only work carried out in the UK has been on a much smaller scale and two studies concerned with relief and control valves are reviewed in this paper. These illustrate the uncertainties associated with calculating failure probabilities under different operating conditions and the wide range of failure rates obtained. The need for a more extensive programme of research and development is clearly recognised and to this end the National Centre for Systems Reliability (with support from the Mechanical Reliability Research Group) has submitted a proposal for substantial funding for work in this area. It is anticipated that this proposal will receive considerable support from both manufacturers and operating companies.

1. INTRODUCTION

The application of the bathtub curve (1) to electronic equipment is fairly straightforward since this type of unit has simple modes of failure such as the short circuit and open circuit conditions. The curve is characterised by a long "random" phase compared to the early and wear out phases. It is economically feasible to remove the early failure portion by running equipment for a few hours prior to despatch (termed soaking or burn in) so that any defective units will breakdown during this period and can be discarded or have faulty components replaced. Electronic systems have now become so cheap that it is more economic to discard failed units than to attempt to repair them and therefore most equipment will be discarded at the first sign of deterioration due to old age and hence it is unlikely that an increasing hazard rate will be detected in operational equipment. This process is accentuated by the rapid obsolence of electronic equipment due to the escalating rate of the development of new technology at the present time. Because most components or subsystems are produced in very large numbers, it is relatively inexpensive to take a realistic sample from the production line in order to test for reliability characteristics. The cost in energy terms is also relatively low. The main operational factors which influence the reliability, such as voltage and temperature, are well established and therefore the reliability of such equipment can be specified in a relatively precise way (2).

It is extremely difficult to define the reliability of mechanical equipment because of the very complex nature of mechanical devices. Not only do they contain a large number of diverse parts but each part may have several modes of failure. Further complications arise since most mechanical units are subject to maintenance and repair which can significantly affect their reliability. Mechanisms which deteriorate by age related processes such as wear, corrosion and fatigue, can be refurbished at suitable time intervals in order to prevent a breakdown occurring with the subsequent high cost of lost production, secondary damage and so on. In spite of the deficiencies in the conditions in which a great deal of maintenance work is carried out, it is normal to assume that overhauled or repaired equipment is as good as new.

2 MECHANICAL VALVES

A mechanical valve is basically a very simple device and as such would be expected to be extremely reliable. It is used in the design of process plants almost without consideration of its reliability and it is possible for large plants to have over 5000 valves of various classifications. In attempting to determine the reliability of valves therefore it is necessary to appreciate the differences between types and this illustrates the difficulties inherent in mechanical reliability assessment.

Valves can be classified by many attributes and the following are some possible suggestions:

a) Function - control, stop, relief,
b) Construction - needle, diaphragm, gate, plug
c) Actuation - solenoid, pneumatic, spring, manual
d) Material - steel, brass, plastic
e) Quality - nuclear, process, domestic
f) Size

It is fairly obvious that if all of these categories are used then even with 10000 valves, the number in each distinct class will be extremely small (only one or two) and therefore of little use for statistical analysis.

An alternative method of classification may be considered more suitable for reliability studies since in this case we are only concerned with the failure of the valve to meet its functional requirements. However it must be appreciated that the other classifications will probably affect the results significantly.

3. CLASSIFICATION OF FAILURE MODES

In order to carry out a realistic assessment of valve reliability, it is first necessary to define carefully the meaning of the term failure. This requires a specification of the function of a valve and the ways in which the individual components can cause a change in these functions.

The primary function of any valve is to CHANGE the flow of material in some predetermined manner by altering a restriction which is in the flow path of the material. The secondary function of any valve is to CONTAIN the material within a predefined boundary.

Subsequently the failure modes can be related to these functional objectives as

a) incorrect flow
b) leakage of material

These modes may be produced by a variety of failure mechanisms related to the elements of the valve as shown in figure 1.

4. CONTROL VALVES

Control valves are used in process control systems in several different operating modes. Here two studies of flow control valves, used in chemical and nuclear environments, are considered to show the effect of high stress on reliability characteristics.

5. CHEMICAL PLANT CONTROL VALVES

The chemical plant from which the first set of failure data was obtained is operated for periods of several months at a time with up to two years between major overhauls. Breakdowns happen quite frequently and the overall availability of the plant is relatively low (<75 per cent). The plant is located in an exposed site within 2 kilometres of the seashore and may be considered to be under higher than normal environment stress. For example there is considerable surface corrosion on the exposed parts of the equipment.

A variety of operating media were involved such as fuel oil, lube oil, organics and aromatics. Temperatures were relatively modest (generally less than 50°C) with pressures in the range 20 to 700 psi. The majority of control valves were operating between 200 and 300 psi and were in the size range 2-6 inches diameter with a median of 3 inches diameter.

Data were made available by the Maintenance Department of the company for a calender period of just over 3 1/2 years on a total population (including replacements) of 394 control valves. The valves were generally of the pneumatically-actuated, globe type with double stuffing boxes and spring-opposed diaphragm actuators. The maintenance History Sheets were quite detailed with over 70 per cent giving a full history of failure including date and time of the incidents, modes of failure, component-part and complete valve replacement details.

Fairly coarse tests were carried out initially on the data using a random sample of 50 valves. From this sample an MTBF of 6890 hours was calculated which showed satisfactory agreement with the overall failure rate of $145f/10^6$ hours obtained from the total population. Most of the subsequent exercises were, therefore, carried out on this sample of 50 valves. However, an analysis of all valve failures was carried out to identify the proportion of failures which fell into each of the main failure modes. The failure mode categories chosen for a previous study (3) were also employed here. The total of 329 failures were reasonably evenly distributed between the seven main failure modes as follows:

Gland Leakage	24%
Seat/Plug Leakage	21%
Body Components	18%
Actuator/Positioner	17%
Seize/Stiff	13%
Blockage	7%
	100%

It was notable that the proportion of failures which could be attributed to degradation of component parts was higher than in other reported analyses(4).

Inspection of these chemical plant data suggest a pattern of early-life failures - over 6 per cent of the total occurred before 10 hours operation after start-up and 14 per cent before 100 hours (4days). This early-life failure pattern was confirmed by Weibull plotting. The Weibull shape factor was 0.4.

Inspection of the failure descriptions on the maintenance History Sheets also showed that the consequences of failure was tending to increase as a function of valve age. Examples were valve plugs snapping off,bolts shearing, seat replacements, replacement of worn valve guides and major repairs on valve actuators.

The analysis showed a mean failure rate for the control valves in this petrochemical plant approximately a factor of 5 higher than the figure of $30f/10^6h$ suggested by Green and Bourne (5) for similar equipment operating in Normal Industrial Conditions. The Weibull Shape Factor of 0.4 associated with times to first failure indicates a failure distribution strongly skewed towards Burn-In (Early-Life) failure characteristics. The implication is that the stresses experienced in this chemical plant led to high failure rates and non-exponential failure characteristics.

6 NUCLEAR PLANT CONTROL VALVES

The control valves employed in the feedwater control systems of a nuclear power station, considered in the second study had been the subject of previous analysis (3). In that study the mean failure rates of process control equipment over the first 7 1/2 years of operation were found to be generally rather lower than predicted by Failure Modes and Effects Analysis.

Control Valves (also the small instrument valves known as Auto-Manual Stations) reached the predicted failure rate after only twelve-months. It is worth noting that control valves and auto-manual stations both had inherent problems. In the case of the control valves the problem resulted from a change in the feedwater throughput due to a permanent reduction in reactor power caused by unsatisfactory materials of construction. The result of this change in operation condition was that the control valves are not oversize for their present duty and operate over a small range of stem movement. This has caused some vibration and significant cavitation damage on the high pressure system valves.

A further visit was made in 1980 to collect new data for this study. Sixteen failures were reported during the previous two year period indicating an overall failure rate for all control valves (HP and LP systems) of 0.5 f/y. This value is in close agreement with the failure rates originally predicted by FMEA (0.52f/y) and observed during the previous period of surveillance (0.48f/y). However, more detailed analysis (6) showed a marked difference between the failure rates of HP and LP control valves (x4.2) with wearout the dominant mechanism on the HP valves where the operating stresses were more severe. The implication is that variations from the design conditions can lead to increased rates of failure. The increases are most marked on the more highly stressed equipment.

7 DISCUSSION OF RESULTS

The detailed results of these studies have been compared with analyses of control valve failures in less hostile operational/environmental conditions. This has led to the conclusion that a constant failure rate model is only suitable for situations where the safety margin is large (Smooth-Loading). Where the safety-margin is small (Rough-Loading), a Chain-Type (Weakest-Link) model appears to give the best representation of the failure characteristics. Two possible developments after repair then appear to be tenable:

1. Weak-links are replaced by strong-links either as a random process of selection from new components with distributed strengths or by redesign or improved manufacture of the weak-link component. In either of these situations the wearout characteristic is unlikely to be exponential and the constant failure rate model will not be appropriate.

2. Weak-links are replaced by weak-links: either by the introduction of components with too low a safety margin or by degradation of components not replaced during repair. In this situation the failure characteristics is unlikely to be exponential and the constant failure rate model will not be appropriate.

Development (1) is clearly an acceptable solution and one in which improved quality control should quickly produce a significant improvement in reliability performance. The second repair situation is a problem when the mean wearout life is low and may be expected to result in high failure rates. Here failure rates considerably higher than the average rates reported in published tables were seen to result in situations where stresses were known to be high. Nevertheless, this is not inconsistent within industry at large - the range of reported failure rates for control valves, in fact, covers a very wide range. Fig. 2 shows the (log-normal) distribution of failure rates reported from air-operated globe valves in the SRS Reliability Data Bank. From Fig. 2 it can be seen that the range of failure rates extend by more than a factor of 10 from lowest to highest with a median value of $20f/10^6h$ (0.18f/year).

It is likely that many of these higher-order failure rates are from control valves where significant degradation is taking place. Detailed analysis of failure data from these plants would probably identify weak components or adverse operating conditions which could be improved at little expense to improve reliability.

8 PRESSURE RELIEF/SAFETY VALVES

The pressure relief or safety valve is an extremely simple mechanical device and most process plants have many of these units. The function of a safety valve is to protect vessels and pipelines from catastrophic failure due to unexpected rises in pressure and it is therefore in a standby mode for the majority of its life. It is very likely that some valves will never be required to operate due to a pressure excursion during the total lifetime of the plant. Nevertheless it is necessary, if the safety valve is to be of any value whatsoever, that it should have extremely high reliability. It is unfortunately difficult to establish the actual times to failure for individual safety valves since there is generally no indication that the valve is unable to function during normal operation of the process. Most companies now remove safety/relief valves periodically and test them on an appropriate test rig in order to determine their performance. They are then usually overhauled and reset before being returned to service. If a valve has been judged to have failed then it is standard maintenance practice to reduce the time before the next inspection.

Whilst the valves which have seized shut or open are obvious failures, in between these two extremes there is an infinite range of possibilities and some guide lines need to be laid down. It is very common to assume that valves which deviate by more than 10% from the cold set pressure have failed and although the selection of this particular value seems to be rather arbitrary it was considered appropriate to use this definition in the initial stage of this work (7).

9 RELIABILITY ASSESSMENT

It was anticipated that safety valves would be tested at very regular intervals with the more important ones being tested every two years whilst those on less hazardous duty might be tested every three or four years. Data collected from the records of several companies showed that this was not the case and only a small number of valves were tested regularly at specified periods when this was necessary to meet safety regulations. The majority of valves were tested after arbitrary periods in service and in some cases were very short. This was sometimes done for process changes or obvious valve failure (feathering or blowing). Some valves had not been tested for a very long time and although these were only used for thermal protection of minor pipelines, this was not considered to be a satisfactory situation if the valve was to be of any real benefit.

Since the valves are tested at arbitrary times, the actual times to failure are not known and the only information available is that, when tested after a given period in service, some valves were found to have failed at some unknown time during this period. Therefore it would be desirable to collect together all valves which had been in service exactly the same time (for example 1 year, 2 years and so on) and establish the proportion of failures in each case. These values could then be used as point estimates of the unreliability. Since it was found that the periods in service did not fall into such simple groups, the time interval was divided on an appropriate basis in order to get approximately 100 results in each group. The proportion of failures would be expected to increase as the period in service increases but Table 1 shows that this is not true. For the total population of valves, the proportion of failures was found to be 44.5% and it can be shown that the variation between each group could be due to sampling rather than an age related phenomenon.

10 EUROPEAN VALVE EXPERIENCE

It has been reported (9) that 7% of all outage times in French nuclear power stations has been caused by valves and these contribute 20% of all incidents. Because of the difficulties of maintaining valves on primary circuits due to contamination, the Electricitie de France (EDF) started a test programme in 1976 which includes the assessment of design and manufacturing followed by testing for conformity to design and performance. The latter includes simple static tests, 1500 cycles of operation under actual service conditions and post test disassembly for detailed examination. Of the 390 valves tested by the end of 1982 only 57% were accepted without modification and 24% were rejected completely. Test facilities at EDF include six closed loop rigs and two open discharge loops. A further two loops are under construction at an estimated cost of £10 million each.

In a study of 1378 performance tests carried out in West Germany between 1970 and 1980 (10) on controlled (pilot-operated) safety valves, it was found that there was a considerable difference between valves which used the process fluid as the control medium and those which used an external fluid (pneumatic or hydraulic). This was attributed mainly to the failure of control elements caused by dirt particles in the control medium. The test programme also indicated a significant improvement in the percentage of valves which performed faultlessly, rising from only 20% of those tested during in the period 1970/74 up to 55% in 1979/80. This improvement was the result of information supplied from the functional tests to manufacturers on weak points and necessary technical improvements. The increased use of reliability analysis and pre-installation testing of components was also considered to have made a significiant contribution to this welcome development.

11 CONCLUSIONS

Mechanical valve failure has always been a source of industrial plant breakdown due to poor value design, incorrect application or bad maintenance practice. However, in recent years, due to the increase in size and complexity of such plant, the dangerous nature of the materials being handled, and the concomitant hazard potential to the operators and/or the community, such failures have been increasingly brought to notice by the national and technical press. Typical examples include:

i) in the USA, the Three Mile Island pressurised water reactor incident began as a small loss of coolant accident caused by the failure of a 5 cm valve to reseat.
ii) in the North Sea the Norwegian Eskofisk-Bravo platform Blowout was caused by a wrongly fitted valve.
iii) in the Process Industries in at least three major accidents, mechanical valve failure was the primary cause.

Evidence provided from accidents in the process industries shows that in over two thousand case histories of major accidents, mechanical valves were found to be the next most frequent cause of failure after pipework. This increasingly alarming record may have resulted from a lack of understanding of reliability attributes which the customer wants and the manufacturer can supply but it is more likely to be the result of buying the 'cheapest' valve, provided it is thought to be 'good enough', without due consideration of subsequent life cycle cost penalties and potential hazards. These factors must contribute significantly to the fact that a considerable share of the UK market is enjoyed by foreign competition. The National Centre for System Reliability, in conjunction with the Mechanical Reliability Research Group, has submitted a proposal to the Commission of European Communities for a substantial grant to investigate this area thoroughly in order to provide British manufactures with the means to improve the reliability of their valves and standards of reliability measurement. This will involve staff of the National Centre for Systems Reliability, university research groups and hopefully some valve manufacturers. We believe that this could be a first step toward the overall improvement in the reliability of British equipment for use in the process and other industries which is vital to our industrial survival.

REFERENCES

(1) Bompass-Smith, J.H., Mechanical Survival, McGraw Hill, London, 1973.

(2) Military Standards Handbook 217B, Reliability prediction of electronic equipment, Sept. 1974.

(3) Moss, T. R., The reliability of pneumatic control equipment: A case study in mechanical reliability, UKAEA NCSR R4.

(4) USNRC, Data summaries of licensee event reports of valves at US commercial nuclear power plant, NUREG/CR-1363.

(5) Green, A.E. and Bourne, A.J., Reliability Technology, Wiley Interscience, 1972.

(6) Moss, T. R., The reliability characteristics of process control equipment, M. Phil thesis, University of Bradford, 1981.

(7) Aird, R. J., Reliability assessment of safety/relief valves, IChemE, vol 60, 1982.

(8) Baby, M., Valve operability in France, SMIRT, 1983.

(9) Oberender, W. and Bung, W., The reliability of controlled safety valves in conventional power plants, 4th National Reliability Conference, paper 4C/1, 1983.

Table 1. Proportion of failures after different periods in service

Period in Service (weeks)	Mean (weeks)	Number of Valves	Number of Failures	Proportion (%)
1 - 39	17.1	104	46	44.2
40 - 57	48.6	104	36	34.6
58 - 90	70.6	107	52	48.6
91 -112	103.0	103	44	42.7
113-147	130.3	103	43	41.7
148-182	160.9	106	51	48.1
183-364	261.0	103	55	53.4
ALL	119.1	746	332	44.5

NB 120 valves had no value recorded for period in service.

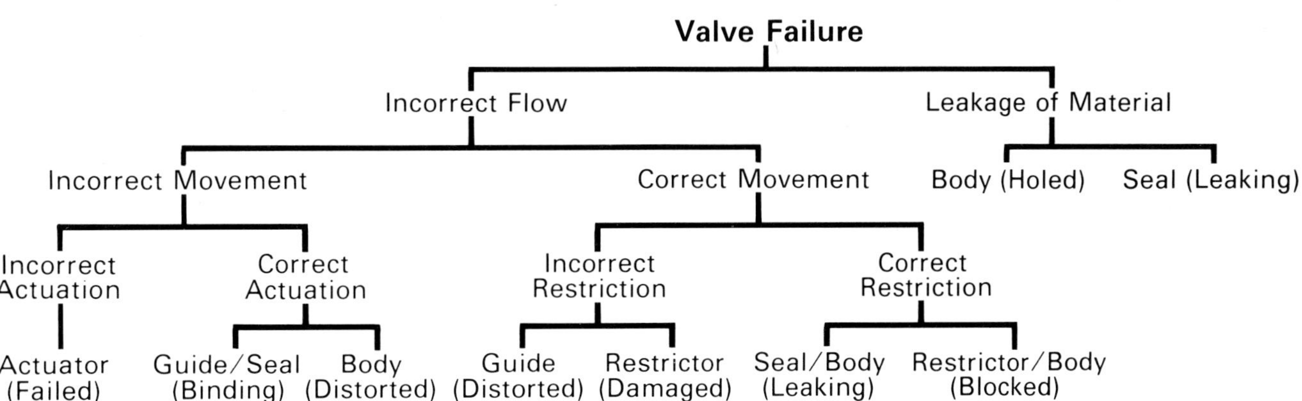

Fig 1 Mechanisms producing valve failure

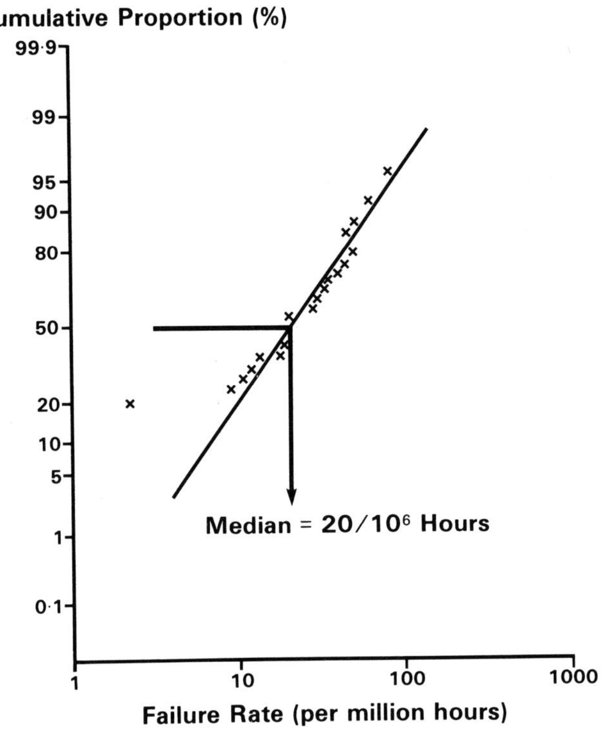

Fig 2 Air operated globe valves — distribution of failure rates

Pump reliability in relation to economized maintenance

R F DE LA MARE, BSc, MBA, PhD, CEng, FIChemE, MBIM
Postgraduate School of Technological Management, University of Bradford, UK

SYNOPSIS
A common conclusion of industrial research carried out by Bradford University is that plant and equipment exhibit a deterioration in reliability performance which can be measured by their Mean Time To Failure (MTTF) order statistics. The results of this paper explain the reasons for these phenomena in relation to a sample of centrifugal pumps. Although optimal pump replacement strategies exist for minimizing Life Cycle Costs, an alternative means of realising economies can be obtained by a combination of better maintenance policies and optimised component replacement.

1. INTRODUCTION

It has long been the philosophy of industrial reliability research, conducted by the University of Bradford, that a sensible analysis of breakdowns and maintenance cost data can point the way to significant economies in maintenance. This research involves a wide variety of equipment ranging from electronic circuits at one extreme, to industrial compressors, pumps, valves, conveyors, crushers and steel rolling mills at the other. Despite this diversity, a common element in the discoveries from this research is that the reliability performance of repaired equipment usually deteriorates between successive repairs and seldom reattains its 'as good as new' condition in a reliability sense.

The purpose of this paper is to explain the reasons for this phenomenon in terms which are useful to the design/maintenance engineer and to explore the economic implications thereof, using the results of a reliability study of pumps for this purpose.

2. THE RESULTS OF PREVIOUS RESEARCH

Previous industrial research conducted at a large chemical factory by Berg (1) showed how wrong it is to treat the failures of similar devices as though they constitute statistically homogenous populations irrespective of their particular repair history. This conclusion was endorsed by Huq (2) and reported in detail by de la Mare (3). In mechanical reliability work it is important to treat the failures of devices with different repair histories separately until they can be proved statistically to come from the same population. The research results shown in Table 1 illustrate the reasons for this concern.

In this table the Breakdown Order Number (N) relates to the numbers of times the homogenous population of pumps had failed, so that N=1 means that they had only failed once from new, whereas N=2 means that they had failed twice and had been repaired and recommissioned once. The important feature of this table is the considerable deterioration in reliability performance as measured by the Mean Time To Fail (MTTF) order statistic μ_N which is defined in the list of symbols. It is important to qualify that the differences in these order statistics were shown by Huq (2) to be statistically significant with high levels of confidence.

A casual observation of the repairs of these pumps suggested that the major reason for this reliability deterioration lay in their method of repair. It was observed that failed parts were renewed as were others which showed visible signs of deterioration but other parts were reassembled without any direct measure of their accumulated damage due to corrosion, erosion, abrasion, stress, fatigue, distortion or creep. It was hypothesised that such a maintenance policy could cause the MTTF order statistics to diminish since un-renewed components would eventually accumulate sufficient damage to exacerbate the failure of other parts as well as failing eventually themselves. A research programme was therefore undertaken to test this idea and some of its more interesting results are contained herein.

3. THE RESULTS OF CURRENT RESEARCH

In mechanical reliability research, one has to balance the cost of the study against the need for adequate sample sizes of data, to allow prediction with confidence, and yet retain sufficiunderstanding of each individual failure to differentiate between subtle differences which might exist between different failure modes and modi operandi. This point is most relevant in the context of centrifugal pumps which possess such a wide range of operating capability that two indentical pumps operated in the same environment but at extreme ends of their capability envelopes could exhibit substantial differences in their reliability attributes. Conscious of this tradeoff, it was decided to monitor the failures of ten pumps which were identical but for a few small differences in their impellor sizes. They were used in pairs, with passive redundancy, to transport chemical products into and away from two continuously operated distillation units connected in series, with the bottom product of the first column being fed to the second. In effect therefore the operating duties of each pump were quite similar but for small differences in their normal operating temperatures. It can be shown statistically that collectively these pumps constitute a homogenous sample. A pump was run until it failed, when it was removed for repair and the standby pump was operated instead.

The problem of how best to measure the reliability of mechanical equipment, which is operated with passive redundancy, is an area of reliability engineering in which there is a dearth of publication and debate. In this study the duration between the date of commissioning the plant or recommissioning a pump and the date of its subsequent failure; henceforth called 'the incremental age between successive failures' was used as the appropriate reliability attribute. It will be appreciated however that other attributes such as the duration between the start or restart of continuous operations of a pump and its subsequent failure, the numbers of pump revolutions or the number of on-off cycles to failure might have provided better measures of reliability, either taken singly or in concert. Unfortunately these other data were unobtainable in the time and with the resources available to this study.

The failtime data listed in Table 2 show the reliability performance of these ten pumps during the study period. In most cases these data were extracted from the log-book which had been maintained meticulously by the work-shop foreman. An examination of these data provides the following provisional results:

(i) There was considerable variability in the times to failure with incremental ages between successive failures ranging from 3 to 1121 days.

(ii) There was a considerable diminution in the mean time to failure order statistic ($\hat{\mu}_N$), especially between breakdown order numbers 1 and 4.

(iii) There was eventually an oscillation in the MTTF order statistics $\hat{\mu}_N$ but on average the pumps never regained their 'as good as new' condition, reliability-wise, as measured by $\hat{\mu}_{N=1}$.

(iv) The inverse coefficient of variation order statistic ($\hat{\beta}_N=\hat{\mu}_N/\hat{\sigma}_N$) gradually approached a value of unity, from an initial value well in excess of unity.

(v) By the end of the study all ten pumps had failed at least eleven times but the number incurring further failures diminished rapidly, as shown by the order statistic m_N. Because of problems associated with statistical confidence when dealing with small censored samples, order statistics are not provided for breakdown order numbers in excess of N = 15.

(vi) There were appreciable differences between the numbers of failures sustained by a pump and its standby. This is most noticeable for Pumps 1 and 2, Pumps 5 and 6 and Pumps 7 and 8 and it suggests that the plant personnel had very definite operating preferences. Despite these differences however, the statistical evidence suggests that these failtime data are homogenous.

(vii) The coefficient of variation ($\hat{\beta}_i$) for the failure of each individual pump approximated unity.

Besides these failtime data, the details of the repair histories of these pumps are known, but limitations of space preclude the full enclosure of those details here. In their stead, two typical repair histories of these pumps are provided by Figures 1 and 2. The repair codes used in those figures are listed in Table 3. A systematic analysis of these illustrations provides the following additional results:

(viii) The initial decline in reliability performance (shown by the dotted lines) of these pumps was quite startling. It should also be noted that this decline is understated because the pumps were actually tested and commissioned on a continuously operating basis well before the plant was put into continuous production, from which date the initial times of failure were calculated. These phenomena were shared by all the pumps.

(ix) Bearing renewal seemed to enhance the pumps' reliability either immediately, if the mechanical seat was fairly new, or belatedly, if the seat was subsequently renewed. This phenomenon is depicted by the thick lines shown on these illustrations. Using these data for all ten pumps it was found that:

a) the probability of the time to failure increasing after the renewal of the bearings was 60% irrespective of the age of the mechanical seat. and

b) the probability of the time to failure increasing after the renewal of the bearing, if the mechanical seat was replaced within 100 days of the new bearings, was 80%.

(x) The time to failure of these pumps declined with the increasing age of their bearings.

(xi) Similar analyses for other pump parts showed similar results to those mentioned in (x) above. In the interests of script economy however they will be published elsewhere.

4. DISCUSSION OF RESULTS

Although the results of the reliability trial outlined above are useful in themselves, reliability engineers often try to develop such results further using mathematical distributions which fit optimally the test data. The object of such an exercise is to improve reliability predictions and to permit a sensible comparison between alternatives, especially when the populations of sample data are small. Several mathematical functions exist for this purpose as do the means for their statistical fitting. It is not our purpose to dwell on such matters here since they are well documented in text books on reliability engineering, such as Mann[4] and BS 5760. However it is relevant to mention briefly one distribution which has been found most useful to this work. This involves the Weibull Distribution, which is described mathematically by the following equation:

$$F(t) = 1 - \exp\left[-\frac{t}{\eta}\right]^{\beta}$$

where the meanings of the symbols are given in Section 8.

Two special features of this distribution are that it plots as a straight line on special graph paper and the value of the shape parameter β can be approximated by the inverse coefficient of variation ($\hat{\mu}/\hat{\sigma}$).

The particular significance of this distribution

to design and maintenance engineers lies in the values which the shape factor 'β' assumes. Two zones of values are particularly relevant, namely:

β < 1 for which off-line preventive maintenance is definitely not justified. This condition applies to a component but not necessarily to a system comprising several components and

β > 1 for which off-line preventive maintenance might be justified depending on its cost effectiveness.

With the assistance of the points mentioned above, we can now reconsider our preliminary results as follows:

Result (i) reminds us that we are dealing with the results of a stochastic process whereby the so-called 'failures' of the pumps occurred randomly due to chance over-stressing of a component part which was usually, but not always, the mechanical seal. This result suggests that the correct interpretation of these failure data requires proper statistical analysis.

Results (ii) and (iii) together suggest that an overall deterioration took place in these pumps, despite their repair. The extent of this deterioration is illustrated by Figure 3 which confirms that the Weibull Distribution fits reasonably the failure data. The large separation of its graphs is but one measure of the confidence one has that these results are statistically different and not just a consequence of chance events. With shape factors well in excess of unity these graphs suggest that the pumps experienced failures due to 'wear-out - old ageing' effects which could possibly have been remedied partially by a policy of preventive off-line maintenance. By contrast the firm pursued a policy of corrective maintenance only.

Results (iii) and (viii) combined lead one to believe that the considerable erosion in reliability was suspected by the maintenance staff who then managed to improve matters partially by accelerating the replacement of bearings: a matter which is discussed in more detail later.

Result (iv) is significant from a reliability viewpoint in that it shows how the particular maintenance policy was eventually causing the pumps to fail exponentially. This phenomenon is well known and understood, and can be explained by the randomness imported into the mechanical device by the particular maintenance policy and the well-known concept 'as bad as old'.

Result (v) reminds us of the risks involved with making decisions on the basis of sparse evidence. It is encouraging however to note that the overall results of this small but in-depth study are endorsed by the much larger studies which were mentioned at the outset of this paper.

Result (vi) reminds us that the meaning of the word 'function', as used in the definition of reliability, often has to be interpreted in its widest sense if the results of a reliability study are to be useful.

Result (vii) is significant from a reliability engineering viewpoint. It means that the failures of one, or more, device(s) considered as a system, without any consideration for their vintage or repair history, is likely to give a completely different reliability signature to that found using order statistics. This means that the reliability attributes of any maintained devices have to be studied carefully.

Result (viii) is another way of showing how the initial deterioration of these pumps, partly brought about by delayed renewal of their original bearings, was made manifest.

Result (ix) and (x) were examined in more detail. Figure 4 shows the distribution of bearing ages at replacement. It should be noted that the pump bearings seldom failed but were replaced according to the judgement of the fitters, without using cardinal measures or standards. In effect these distributions comprise the fitters' subjective estimates of the failtime distributions for the bearings if they had been left to fail belatedly. The significant feature of these distributions is their age dependency as indicated by values of β well in excess of unity. This phenomenon was tested by other means as shown in Figure 5, which demonstrates clearly the strong inverse relationship between increasing bearing age and pump time to failure. This result must now be examined in more detail to determine whether more frequent bearing replacements would be justified economically. This result also suggests that a judicious application of condition monitoring techniques (such as vibration analysis) applied to the bearings in particular might be made cost effective.

Result (xi) is further proof that an overall process of deterioration was occurring which might be remedied economically through the proper application of preventive maintenance routines.

5. THE ECONOMIC IMPLICATIONS OF THE RESULTS

Previous work by Berg (1) showed that a policy of regularly replacing these pumps by brand-new ones would realize substantial savings. Based on MTTF statistics for successive ordered failures and estimates of capital and operating costs, he showed how unique optimal replacement strategies existed for each type of plant operated by this company, including the pumps mentioned already. These replacement strategies were optimal in their minimizing the Life Cycle Costs (LCC) incurred with the purchase and maintenance of these pumps. Some of his results were shown in Table 4. Furthermore he proved that these optimal strategies were robust and relatively insensitive to errors and changes in the economic input data. These results do not include the potential revenue to be realized by having the pumps on-stream for an estimated extra one percent of time. It is known that if the firm could sell this extra production then the revenues so realized would far exceed the maintenance savings made by the more efficient use of these pumps. In effect therefore, these economic benefits of optimized replacement are understated and the composite weighted saving of 20%, corresponding to the pumps and valves mentioned in Table 4, is conservatively low.

The results of current research endorse Berg's conclusions, namely that optimal replacement strategies exist. They also provide some of the reasons for the MTTF deterioration and suggest a means to minimize this effect through the judicious application of more appropriate maintenance policies and procedures; such as the more timely replacement of bearings, condition monitoring and the scientific use of measures and standards, as they relate to machine wear. The ecomomic merits of these alternatives are currently being assessed.

6. CONCLUSION

This case study shows how a formal collection of data relating to plant and equipment breakdown followed by a statistical analysis of the results, can point the way to significant cost savings. Within the context of this research programme and according to the maintenance policies and procedures of the firm involved with this research, these savings are made possible by the timely replacement of these pumps in the face of a deterioration in their MTTF order statistics. This paper indicates the reasons for this deterioration and suggests that greater savings could possibly accrue from a more scientific approach to maintenance management, possibly by using preventive maintenance coupled with condition monitoring to optimize collectively the repair and the subsequent replacement of these pumps.

7. REFERENCES

(1) BERG, Ø. 'The Terotechnological Implications of Capital Plant Breakdown'. Unpublished PhD Thesis. University of Bradford, 1977.

(2) HUQ, M D, NUREL. 'Process Plant Failure' Unpublished MSc Thesis, University of Bradford, 1978.

(3) DE LA MARE, R F. 'Optimal Replacement Policies'. National Centre of Systems Reliability Research Report NCSR R21 (available from National Lending Library but also from the author). 1979.

(4) MANN, N R, SCHAFER, R E and SINGPURWALLA, N R. Methods for Statistical Analysis of Reliability and Life Data . Wiley, New York, 1974.

8. SYMBOLS

β = shape parameter of the Weibull distribution (dimensionless).

exp = exponent and base of the Natural Logarithm.

$F(t)$ = cummulative probability of failure before t.

i = pump number.

m_i = numbers of failures of the ith pump.

m_N = number of pumps incurring N successive failures.

MTTF = Mean Time to Fail = $\hat{\mu}_i$ or $\hat{\mu}_N$ (days).

n = number of pumps tested

N = breakdown order number : the number of successive failures experienced by a pump

η = characteristic life parameter of the Weibull distribution (days).

$\hat{\rho}_i = \hat{\mu}_i/\hat{\sigma}_i$.

$\hat{\rho}_N = \hat{\mu}_N/\hat{\sigma}_N$.

$\hat{\sigma}_i$ = variance of the incremental age to failure of the ith pump (days^2).

t = failtime = $t_{i,m}$ or $t_{i,N}$ (days).

$t_{i,m}$ = incremental age to failure of the ith pump between its (m-1)th recommission and its mth failure (days).

$t_{i,N}$ = incremental age to failure of the ith pump between its (n-1)th recommission and its Nth failure (days).

$\hat{\mu}_i$ = MTTF for the ith pump.

$$= \frac{t_{i,1} + t_{i,2} + + t_{i,m} + \ldots t_{i,m_i}}{m_i}$$

$\hat{\mu}_N$ = MTTF for all pumps incurring their Nth breakdown only.

$$= \frac{\sum_{i=1}^{n} t_{i,N}}{m_N}$$

$\hat{\sigma}_N$ = variance of the incremental age to failure of all pumps incurring their Nth breakdown only. (days^2).

TABLE 1. Examples of the MTTF order statistics for centrifugal pumps subject to repair.[1]

Pump	Pump A1		Pump A2		Pump B1	
Breakdown Order Number (N)	Sample Size(m_N)	$\hat{\mu}_N$	Sample Size(m_N)	$\hat{\mu}_N$	Sample Size(m_N)	$\hat{\mu}_N$
1	54	387	25	406	15	362
2	50	213	23	397	15	271
3	40	207	21	307	14	141
4	39	223	20	253	13	255
5	36	183	18	210	12	344
6	31	149	15	243	10	375
7	30	235	9	211	9	187
8	28	255	8	139	9	167
9	26	198			7	145
10	22	151			7	157

Note 1: All the order statistics $\hat{\mu}_N$ allow for the effects of censoring.

TABLE 3. THE REPLACEMENT-REPAIR CODE FOR THE STILL-PUMP REPAIRS

P	=	MECHANICAL SEAL PARTS
S	=	IMPELLOR SHAFT
S1	=	MECHANICAL SEAL (COMPLETE)
St	=	SEAT
B	=	BEARINGS
O	=	OIL-SEALS
I	=	IMPELLOR
RS	=	RESPRAYED SHAFT
M	=	MISCELLANEOUS

TABLE 2. DISTILLATION-UNIT PUMP FAILURES:
INCREMENTAL AGE (DAYS) BETWEEN SUCCESSIVE BREAKDOWNS

Breakdown Order Number (N)	Pump Number										Order Statistics			
	1	2	3	4	5	6	7	8	9	10	$\hat{\mu}_N$	$\hat{\sigma}_N$	$\hat{\rho}_N$	m_N
1	228	441	415	787	112	394	110	411	291	138	338	206	1.61	10
2	539	368	109	302	438	161	286	89	233	160	269	147	1.82	10
3	225	244	9	291	201	196	29	28	407	631	226	191	1.18	10
4	39	217	141	98	167	21	40	56	72	74	93	63	1.46	10
5	31	87	198	752	12	180	69	50	21	91	149	221	0.67	10
6	39	89	67	137	218	21	17	147	12	71	82	67	1.21	10
7	11	130	9	173	33	199	37	27	130	137	89	72	1.23	10
8	44	74	33	165	95	116	23	79	75	246	95	67	1.41	10
9	32	19	634	335	127	38	15	23	61	148	143	198	0.72	10
10	63	227	404	43	201	96	15	67	394	342	185	151	1.23	10
11	100	416	137	13	48	4	37	16	458	16	125	170	0.73	10
12	71	727	316		1121	65	3	33	46	19	267	396	0.67	9
13	263	131	139		227	152	74	144	25	119	142	72	1.98	9
14	8		98			30	15	156	103	16	61	58	1.05	7
15	61		400			549	16	560	89	146	260	236	1.10	7
16/26	281					107	33/31		240	70				5
17/27	116					76	40/28		20	145				5
18/28	100					42	34/57		43	88				5
19/29	217					181	18/6		29	60				5
20/30	24					31	25/15		178	174				5
21/31	331					47	23/68		83	96				5
22/32	46					190	48/207		30	53				5
23/33	12					39	26/160		35	138				5
24/34							69/61			54				2
25/35							28/280			33				2
$\hat{\mu}_i$	125	244	207	281	231	128	58	126	134	131	INDIVIDUAL			
$\hat{\sigma}_i$	134	198	185	263	289	129	70	156	137	128	PUMP			
$\hat{\rho}_i$	0.93	1.23	1.12	1.07	0.80	1.0	0.83	0.81	0.98	1.02				
m_i	23	13	15	11	13	23	35	15	23	25	STATISTICS			

TABLE 4. SAVINGS TO BE REALIZED FROM OPTIMAL REPLACEMENT STRATEGIES

Equipment and Type	Optimal Number of Repairs	Overall Savings to be realized by optimal replacement (%)
Pump A1	8	21.7
Pump A2	4	14.1
Pump B1	6	17.6
Pump C1	5	9.7
Pump D1	1	26.4
Valve V1	5	14.8
Valve V2	5	21.1

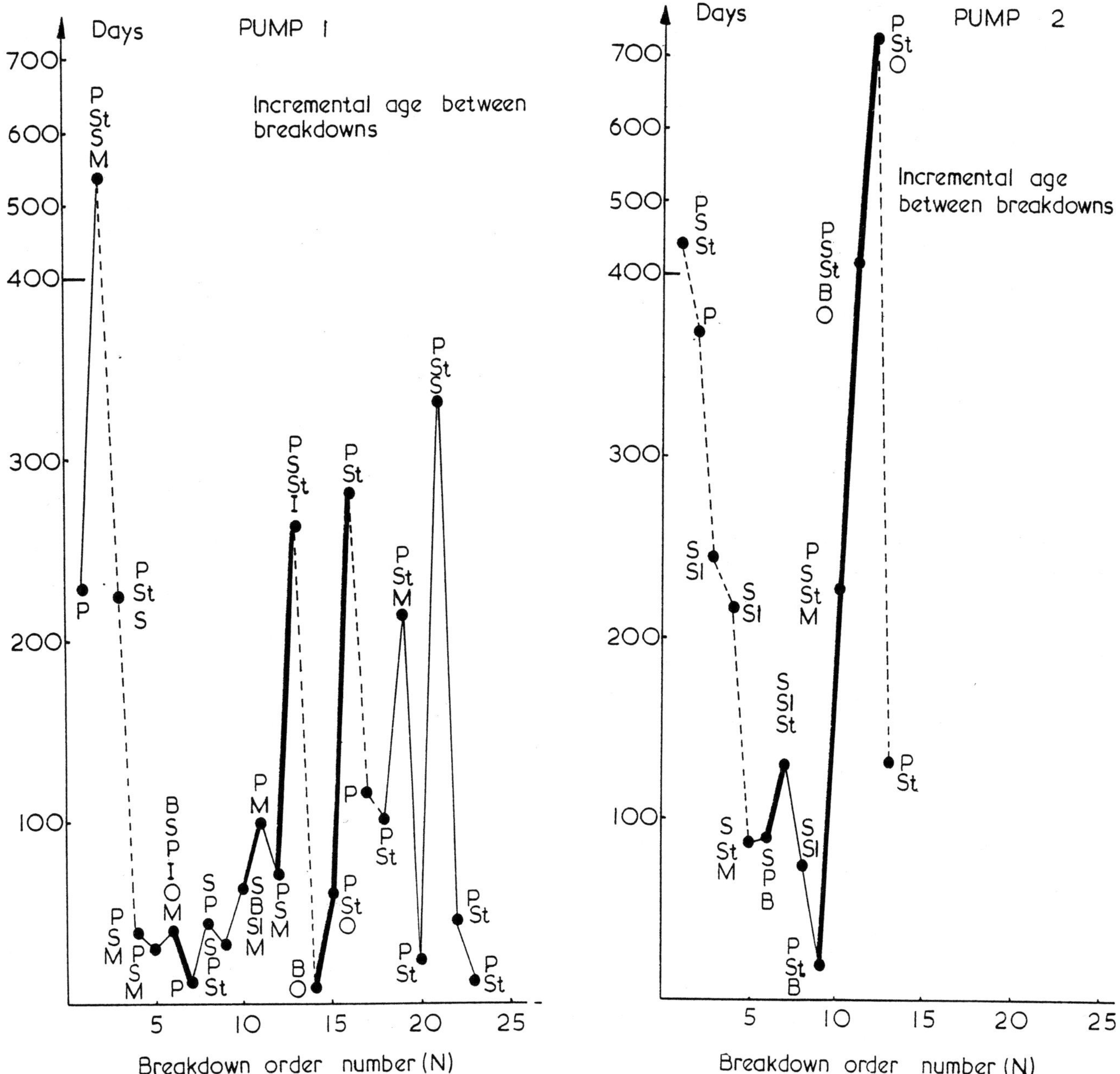

Fig 1 Failtime and repair data for Pump 1

Fig 2 Failtime and repair data for Pump 2

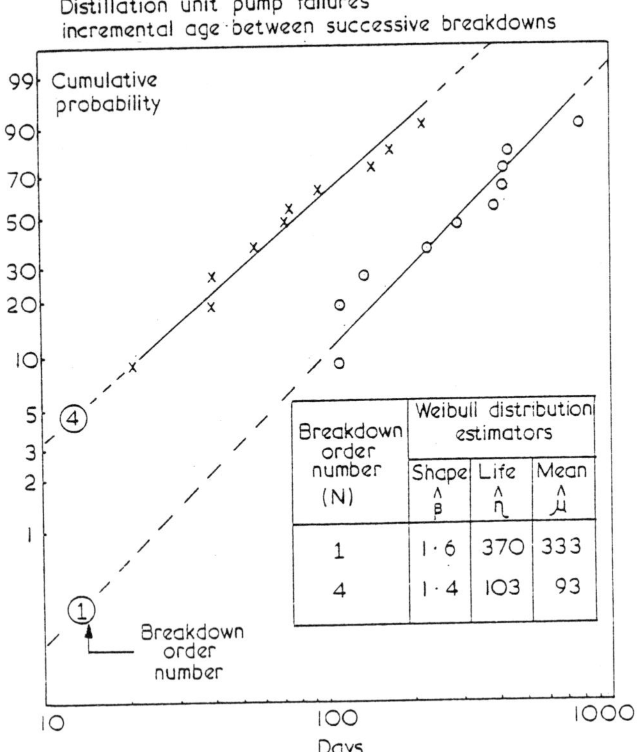

Breakdown order number (N)	Weibull distribution estimators		
	Shape $\hat{\beta}$	Life $\hat{\eta}$	Mean $\hat{\mu}$
1	1·6	370	333
4	1·4	103	93

Fig 3 Weibull distribution plots for pump ordered breakdowns

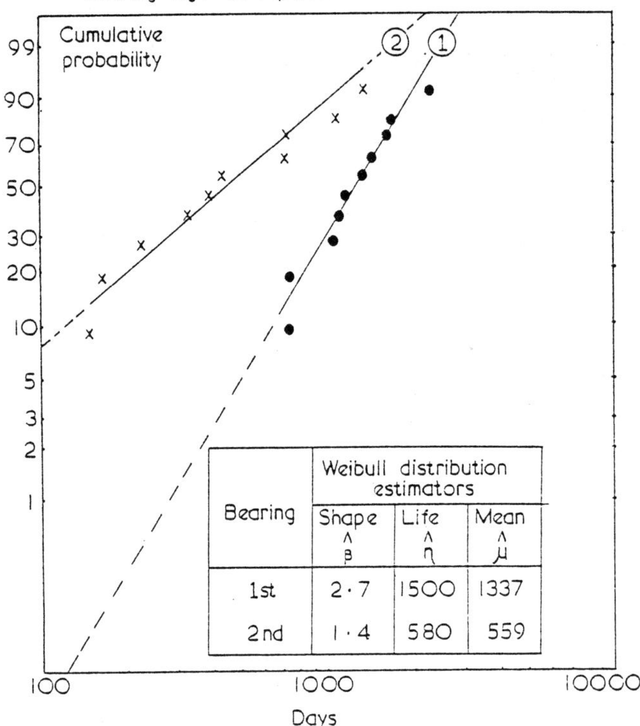

Bearing	Weibull distribution estimators		
	Shape $\hat{\beta}$	Life $\hat{\eta}$	Mean $\hat{\mu}$
1st	2·7	1500	1337
2nd	1·4	580	559

Fig 4 Weibull distribution plots for bearing ordered replacements

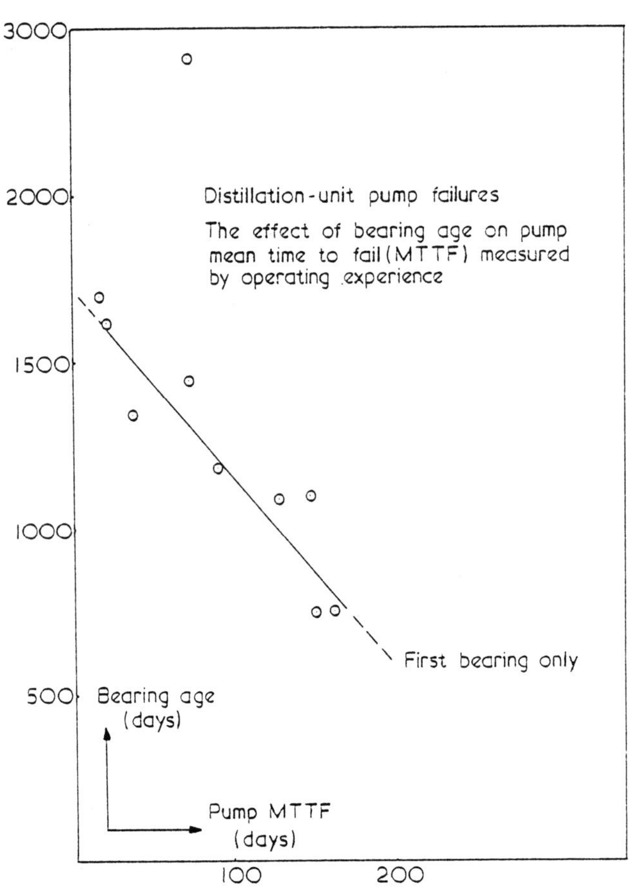

Fig 5 The effect of bearing age on pump rehability

Implementation of a reliability programme on a new gas compression station – A case history

T P LITTLEJOHNS, MSc and M KELLY, BSc
British Gas Corporation

SYNOPSIS When selecting machinery for compressor stations the British Gas Corporation seeks to ensure that in operation the equipment will achieve the required performance, safely, economically and consistently. Standard reliability engineering techniques are used to quantify reliability in terms of start probability and mean running time between failures (MTBF).

For one of its recently constructed gas compressor stations a newly developed compressor was selected as part of the machinery package. This station was also the first station on which reliability targets were set as part of the specification. The targets were set at the level of the best performance of units then in service. To ensure these targets were achieved within a stated operational period a reliability programme was initiated.

Any new design, unproven in operation, can carry a greater than normal risk of experiencing problems and therefore giving unreliable operation. For this compressor these complications materialised. This paper describes the problems and how they influenced unit reliability. Although the paper confirms that standard statistically based reliability engineering techniques cannot adequately warn of problems in new designs, it suggests that a reliability trial can assist problem investigation and solution.

1 INTRODUCTION

The British Gas Corporation operates and controls a complex gas transmission network designed to ensure security of gas supply, in a safe, reliable and efficient manner to meet widely varying summer and winter demand. The network is an integrated, countrywide, system of pipelines, which receives natural gas at shore terminals and transports it at high pressure to regional offtakes for distribution at a low pressure to the customer. At a number of carefully selected points along the high pressure system, a series of aero derivative gas turbine powered compressors units have been installed to increase the flow capacity of the pipeline. At present there are 15 compressor stations, comprising 44 operational machinery units.

Each machinery unit is installed in its own building and consists of a train of three sub-units, namely, a gas generator, a free power turbine and a centrifugal compressor as shown in Figure 1.

The sub-units, with their support systems are mounted on to one or more bedplates connected to form a single substantial structure. The main supporting systems required to operate and monitor the machinery include fuel gas, lubrication and sealing oil equipment, as well as instrumentation for control and protection. High velocity ventilation ensures a safe environment within the building and the air intake and gas exhaust systems are acoustically designed to maintain acceptable external noise levels.

When selecting machinery we seek to ensure that in operation the equipment will achieve the required performance safely, economically and reliably. We use standard reliability engineering techniques to quantify reliability in terms of start probability and mean running time between failures (MTBF).

2 RELIABILITY ASSESSMENT OF NEW PLANT

Within British Gas it is now common practice to include reliability assessments in plant evaluations. The two most frequently used methods for assessing reliability at the specification and design stage are the Generic Part Count and the Fault Tree Analysis:

(a) Generic Part Count Technique

The Part Count method predicts the reliability characteristics of the overall equipment from consideration of its detailed parts. The depth of the assessment, is determined by the detail of the reliability information available.

If there is sufficient information available the reliability assessment will consider the failure modes of each part and their effects on the overall equipment performance. This is termed a Failure Modes and Effects Analysis (FMEA).

(b) Fault Tree Analysis

The analysis proceeds by identifying the potential failure situations. The chains of events that lead to these situations are developed, and the reliability of the equipment is predicted from consideration of the probabilities that the combination of events will occur.

The approach chosen will depend on the problem and the data available.

The information required for reliability assessments is generally based on previous experience and obtained from one or more of the following sources:

(i) In-house records.

(ii) Co-operation with other plant users who share common experience.

(iii) Commercially available data bases. (E.g. the United Kingdom Atomic Energy Authority (UKAEA) Systems Reliability Service.)

(iv) Published sources.

It is generally possible to obtain a satisfactory reliability assessment in the absence of complete data because the majority of problems arise on a minority of components. Concentrating on these dominant components will usually yield a reasonable assessment.

The assessment is primarily directed towards forecasting the rate of failures that may be expected to arise from operation within the design envelope, due to randomly distributed failures of components. It assumes that:

(a) The equipment is capable of performing its required duty as designed.

(b) There are no latent design problems to consistently cause failure.

The second assumption is critical to the prediction and, as will be shown in this paper, can lead to significant differences between forecast and actual reliability performance.

3 RELIABILITY PROGRAMME FOR A NEW COMPRESSOR STATION

3.1 The Machinery Selection

In general we aim to select well proven equipment. However, for one of our recent two-unit compressor stations, we considered a number of new features and requirements, which called for certain innovations in the type of machinery and supporting equipment selected. The question of reliability was therefore brought to the forefront.

The station was the first within British Gas to have the capability of running unmanned with operation being controlled from a remote centre. The telemetry system needed to support the remote control was fed by a mini-computer based machinery control system. Because of the remote operation, starting reliability was considered to be of paramount importance.

Although the station had a requirement for two machinery units, we proposed to save considerable costs by utilising an existing refurbished gas generator for one unit. Therefore, it was important to ensure that the reliability of the older machine, when installed and interfaced with new equipment, also met the Reliability Specification.

The machinery selection was also influenced significantly by the future operating costs. We wanted the maximum available efficiency from the centrifugal compressor to minimise the payback period. A new standard of higher efficiency machines was available, but improved aerodynamics were often incorporated with other new or updated design features. Reliability assessment of the compressor was therefore difficult because little operational experience had been achieved and hence the data available was of limited relevance.

3.2 Reliability Targets

The following targets were set at the level of the best performance of units then in service, and considered achievable without requiring any additional research and development:

(a) To achieve a start reliability of 0.95 by the end of 200 starts after acceptance.

(b) To achieve a run reliability giving a Mean Time Between Failures of 475 hours by the end of 3000 running hours after acceptance.

To ensure the targets were achieved within the stated operational period a reliability programme was initiated. The programme was a co-operative venture with the supplier, but one in which the onus for the achievement of the specification rested with the supplier. The programme consisted of three phases:

(a) A Design Assessment to identify potential problem areas and, if possible, eliminate these prior to manufacture and installation.

(b) Works Test to confirm functional operation.

(c) An Operational Trial to evaluate reliability and identify and correct problems to achieve the specified targets.

The programme was managed by a joint British Gas Corporation/Supplier Reliability Committee. During the Operational Trial members of this committee met periodically to review the operational data and allocate, by agreement, failures to the Supplier's equipment.

4 DESIGN ASSESSMENT

The design assessment consisted of:

(a) A reliability prediction,

(b) The identification of problems on similar units already in service.

The reliability prediction employed the part count technique and drew on the most applicable published data for failure rates. It was supported by limited Failure Modes and Effects Analysis in some areas. Of necessity the prediction contained a number of simplifying assumptions. In particular that failures would occur randomly and that all major components would operate without latent design problems.

The analyses predicted that the start reliability target would be met, ie, the machines would provide a 95% start reliability. The run reliability targets were considered more difficult to achieve. However, no single area was identified as the most significant influence.

Reliability data from units in service at the start of the design assessment was limited. However, as relevant information became available it generally supported the results of the reliability prediction. In addition it highlighted a number of minor potential problem areas and corrective action was implemented.

The reliability prediction and review of inservice problems generated the following changes to the proposed design:

(a) Minor modifications to the seal oil control and lube oil control systems.

(b) Changes of components within the fuel gas and power supply systems.

(c) A reduction in the number of shutdown protection systems (ie, trips). A total of 10 trips were deleted or downgraded to alarms.

(d) The installation of a dedicated chart recorder for start failure fault diagnosis.

In addition there was an increase in the scope and depth of Quality Assurance checks to ensure the fitness of critical equipment, especially where problems had occured on earlier units. The oil systems, electronics and instrumentation were the areas where most of the additional effort was applied.

5 PRELIMINARY OBSERVATIONS FROM THE WORKS TESTS

5.1 The Gas Generator and Power Turbine

The gas generator and power turbine were successfully tested by the main supplier. Despite rigorous control system checks, the tests gave no cause for concern regarding the validity of the reliability assessment. However, we were still far from operating the machines under true site conditions.

5.2 The Compressor

The centrifugal compressor tests were conducted separately from the gas generator and power turbine. Preliminary results from the compressor tests indicated that the machines were not running in a satisfactory manner. Performance efficiency was less than predicted, and mechanical problems were resulting in excessive vibrations.

After a number of visits to the vendor it became obvious that the performance defficiency could be solved, whereas the mechanical problems were fundamental to the design of the machine. Thus we were entering a development investigation of new equipment and the assumptions of the original reliability assessment were no longer valid.

5.3 The Compressor Problems

The basis of the mechanical problems lay within a combination of shortcomings in the seal oil system design and the compressor rotor dynamics. The seal elements were not working correctly but their problems were being compounded by a shaft system instability.

The seal system is shown in Figure 2. The inner carbon face seal is on the high pressure gas side and the two outer floating bushing rings seal to atmosphere. The space between inner and outer seals is filled with high pressure oil, controlled to maintain a positive oil to gas pressure across the face seal. The gas pressure can vary from 40 bar up to 70 bar. This arrangement is needed at both ends of the shaft adjacent to each bearing assembly.

Oil pressure control is vital. Excessive pressure will push oil past the face seal and flood the compressor. Too little pressure will allow gas into the seal and bearing assemblies, resulting in a highly volatile oil and gas mixture passing to other parts of the machinery, including the hot turbine.

The floating bushing seals were at the root of the problem. Due to the high differential pressure across each ring they were locking up within the seal housing but only under certain unknown conditions. The loading forces across one bushing ring is shown in Figure 3a. Their lack of ability to float and move to cater for any small shaft movements was having a detrimental effect on the shaft vibration characteristics. It appeared that the rotor shaft system was not always being simply supported by the radial bearings but was being constrained by the two sets of bushing rings. In effect the bushing rings were acting as another set of bearings at each end of the shaft, producing different and detrimental shaft dynamics.

The first action taken was to off-load the high axial thrust on the bushing rings. This was achieved by reprofiling part of the ring and bleeding high pressure oil to the rear face. The resulting pressure profile is shown in Figure 3b. The success of the modification was only partial. The anticipated reduction in vibration levels was not sufficient to meet the pass-off limits.

The second action was to remove one of the two bushing rings. The result was dramatic. The high vibration reduced to very low and acceptable levels, but although the remaining seal performed well, the seal oil leakage rate increased more than predicted. This latter aspect gave some cause for concern because we were now using most of the available capacity of the oil supply pumps. Nevertheless, it was considered we had overcome our primary vibration problem and, provided the compressor and pumps exhibited similar characteristics under site conditions, our solution would remain valid.

6 THE OPERATIONAL TRIAL

6.1 Reliability Data Logging

In order to allow for a debugging and "running-in" process, we had specified that the first 20 starts and 200 hours of operation should be discounted in terms of the reliability trial.

Once the trial had commenced we logged all the unscheduled machine shutdowns or "trips". For each trip, a detailed job card was made out, which recorded all relevant information available from the unit control monitoring system. Such information relevant to a start failure would include the start recorder charts, showing fuel gas pressure, gas generator speed and turbine temperature. A running trip would be accompanied by our computer data logging record of all other temperatures, pressures, flows and speeds. By this means we were able to build up a picture of the events leading to a failure and embark on a course of action.

Each trip attributed to the machinery was then recorded to show how the reliability was proceeding in relation to the targets. The cusum chart (1), (2), (3), was used to provide a visual indication of our progress. Figures 4 and 5 show the start and run reliability plots for one machine. The reliability of the unit between any two points is indicated by the slope of the line joining them. The chart is arranged so that a horizontal line is on target, an upward slope is better than target and a downward slope is worse than target.

6.2 In Service Result

The compressor vibration levels were maintained at an acceptably low level, but the main seal oil supply pump, being mechanically driven from the power turbine shaft, was unable to satisfy the greater flow requirement at low speeds. We were forced to introduce an operational restriction to avoid the low speed range, until a suitable pump could be identified and installed.

Our cusum charts for the trial (Figures 4 and 5) show that initially the start reliability plots were close to target. However, after about 50 starts the start reliability of both machines began to deteriorate. The run reliability started well below target to the extent that it was very soon realised we would have great difficulty in achieving the target figures within the operational trial period.

Analysis of the particular trips causing these reductions indicated two problem areas. The fuel gas control system was leading to inconsistent ignition and light-off during the gas generator startup. Under running conditions the compressor lubrication and seal oil system was suffering transient pressure dips, especially when accelerating to full speed.

Comparison of our start records from the chart recorders revealed that fuel gas ignition did not always occur at the same gas generator speed. The period between "ignitors-on" and "light-off" also varied and no obvious trend difference existed between the new and refurbished gas generators. The fuel control system rather than the gas generators was believed to be the source of the problem. Analysis of the chart recorder traces highlighted a number of changes necessary in the control parameters and the way in which control was being exercised.

The compressor lubrication and seal oil system provided the major contribution to the low running reliability. Initially we believed that our original bushing ring problem was causing pressure fluctuations due to difficulties in controlling the higher oil flow rates. To some extent this was true, but we also highlighted another underlying problem associated with the supply pump control. It was stated earlier that the main supply pump was mechanically driven from the power turbine. Because the output of this type of pump is proportional to engine speed, it is incapable of delivering sufficient flow at low speeds. To overcome this dif-ficiency, the design had included an electrically driven auxilliary pump to cover the startup phase, which automatically switched off after a certain speed and allowed the main pump to take over supply. Despite a considerable overlap between the two pumps, the system was experiencing a large pressure drop as the electric pump shutdown. Detailed investigations of flows, pressures and the control of spill back valves eventually provided an answer, and changes to pipework and valves were required.

6.3 Presentation of Results

Throughout the Operational Trial, the machinery supplier participated in the exercise of recording and analysing data. The factual aspect of the resulting reliability figures and the obvious trends produced by the cusum charts, served as a permanent indicator of performance, and amplified the need to improve matters. With the aid of the reliability programme we were able to demonstrate the persistence of certain problem areas, and the inconvenience and cost of an unreliable machine. It highlighted to the compressor vendor, the dominance of the large number of seal oil system control trips in comparison to the rest of the machinery train.

It was perhpas this latter aspect that also encouraged him to produce an improvement to the shaft dynamics. By introducing an improved bearing design, we were able to return to the original two bushing ring arrangement, with the lower oil flow requirement. This in turn avoided the need to instal larger pumps and produced a much improved and consistent seal oil control.

The cusum chart for run reliability, (Figure 5) indicates the cumulative effect of these modifications with a significant improvement after 600 hours.

7 REVIEW OF RESULTS

7.1 The Design Assessment

Predictions for the start reliability provided a good forecast, with both units meeting their target figures by the end of the operational trial. Greater concentration on achieving the correct starting logic and fuel gas setting-up procedures probably helped to achieve this.

The final run reliability figures were significantly short of the predictions. The changes recommended by the design assessment were of little consequence in terms of the eventual overall reliability, although component improvements based on experience from other compressor stations were successful.

The difficulty in assessing the small improvements was masked by the more major design problems. However, it is more constructive to state that the design assessment did not produce more unreliability by its recommended changes and verified the existing designs where-ever any meaningful investigation was conducted. It is an important fact that throughout the operational trial we suffered few unexpected component failures, despite the many fundamental design problems which imposed a more arduous operational condition on many of the systems.

We established a number of areas where stricter quality control had to be exercised, but any true improvements gained are difficult to quantify.

7.2 The Operational Trial

Despite the initial good start reliability the figure soon reduced to 0.85 over the next 100 starts. We experienced inconsistent starting and instituted improvements in the logic control and hardware associated with the fuel gas system. At the end of the trial we concluded that both machines had effectively reached the target of 0.95.

The run reliability, as stated earlier failed to achieve the predicted targets due to the mechanical design and control weaknesses of the compressor. The degree of unreliability however must be shown in the correct context.

Detailed analysis showed that most trips occured either after initial start up or just before shutdown during a pump changeover phase. Because our operating pattern was more transient than expected, especially with the large number of short test runs required, we recorded a high number of failures over relatively few running hours. This yielded an exceptionally low Mean Time Between Failures. A requirement for long duration runs would have produced very different results.

Another distorting factor was that the compressor created a more arduous set of operating conditions than expected for many of the associated components and systems. Not only were we limited in operational terms to avoid certain known problem areas, we also conducted many unsuccessful tests and introduced minor modifications which in turn imposed a strain on other equipment. Failed test runs produced more shutdown trips, and changes to the seal oil system often adversely affected the lubrication system and vice-versa. It therefore may be argued that we often logged more failures than were strictly due to the original equipment design.

Throughout the programme we were continuously attracted to the possibilities of subtracting the compressor figures from the final data. We had adopted this process for revealing secondary problem areas and developing a priority listing for other investigations. It soon became apparent we could also use this technique for presenting a much improved, and perhaps more accurate reliability picture, for the rest of the machinery. Without the compressor problem the MTBFs of the units are in excess of the target value, 600 hrs for one unit and greater than 800 hrs for the other. Unfortunately it had the effect of distracting our efforts from the principle compressor problem, having derived a false sense of achievement. We therefore continued to report overall reliability and show the machinery package in its worst light.

7.3 The Achieved Reliability

At the end of the Operational Trial the reliability values read from the cusum charts were:

	Number of Starts	Start Reliability	Hours Run	Run Reliability (MTBF)
A unit	270	.94	1600	160 hrs
B unit	274	.98	2271	120 hrs

The start reliability of A unit is considered acceptably close to the tartet.

The run reliabilities at the end of the trial were totally determined by the compressor problems.

8 CONCLUSIONS

The reliability prediction did not match the in-service running performance over the operational trial in terms of Mean Time Between Failures. Its inability to do so was not directly the result of a poor reliability assessment on the area considered, but due to the assumption that the basic design was correct and that design errors if any could be identified and corrected during the early stages of the trial. Our assumption may have been realistic had it not been for the compressor problems. With the advantage of hindsight we might have amended our assessment and target figures to reflect the use of new equipment. There is no doubt that the compressors severly reduced the reliability of the machinery train, but it would be incorrect to conclude that the reliability assessment and its forecast were meaningless.

Initially the compressor accounted for 60 to 70% of the unit unreliability. However, there was still appreciable debugging of minor design problems on the rest of the unit at this stage. By the end of the trial the compressor was accounting for virtually 100% of the unreliability.

A realistic operating trial must continue for an adequate number of hours in order to derive a satisfactory assessment. Experience gained during the operational trial indicates that debugging minor design problems can take up to 500 hrs. We had selected 3000 hours for each unit over a period of one year, but actually only achieved 1500 hrs on one machine and 2000 hrs on the other in 27 months. Much of our running was restricted to short periods which produced a very low estimate of run reliability.

Despite the transient nature of running, the principal benefits derived from the reliability programme arose during the operational trial, where clear identification of problems often led to immediate corrective action. Whilst design effort was concentrated on the compressor, the more detailed analysis of operational data identified underlying trends and secondary problem areas, which were often more amenable to swift remedial action. The implementation of the start recorders and data logging system can be immensely beneficial tools in the fault diagnosis process.

The final and overriding conclusion is that fundamental design problems on recently devloped equipment can dominate the reliability of the complete machinery system. To highlight and overcome this aspect in the reliability assessment stage, even if possible, would require a far more detailed review of critical and unproven areas than a statistically based reliability prediction.

9 THE FUTURE

The reliability programme has formed the basis for a similar exercise on the latest compressor station project. We are now able to avoid some of the less productive areas in the initial design assessment and set more realistic targets for the operational trial based on the machinery selected and its operational profile.

The use of fault diagnosis equipment is becoming a standard on new machinery. The technology available is moving swiftly with the use of microporcessors and we are having to tread cautiously, but the value of on line information retrieval is essential for problem diagnosis and correction.

In parallel with the reliability programmes British Gas has evolved a sophisticated method of data collection which produces a fast identification of in-service problems. The value of swift and accurate fault diagnosis is improving the standard of this data and the collection process. Such information is necessary not only within the Corporation, but can also assist suppliers to respond to particular design problem areas.

REFERENCES

(1) BS 5703 Data Analysis and Quality Control Using Cusum Techniques.

(2) Bissel, A.F. An Introduction to Cusum Charts, Institute of Statisticians.

(3) Van Dobben de Bruyne, C.S. Cumulative Sum Tests. Theory and Practice. Charles Griffin, London.

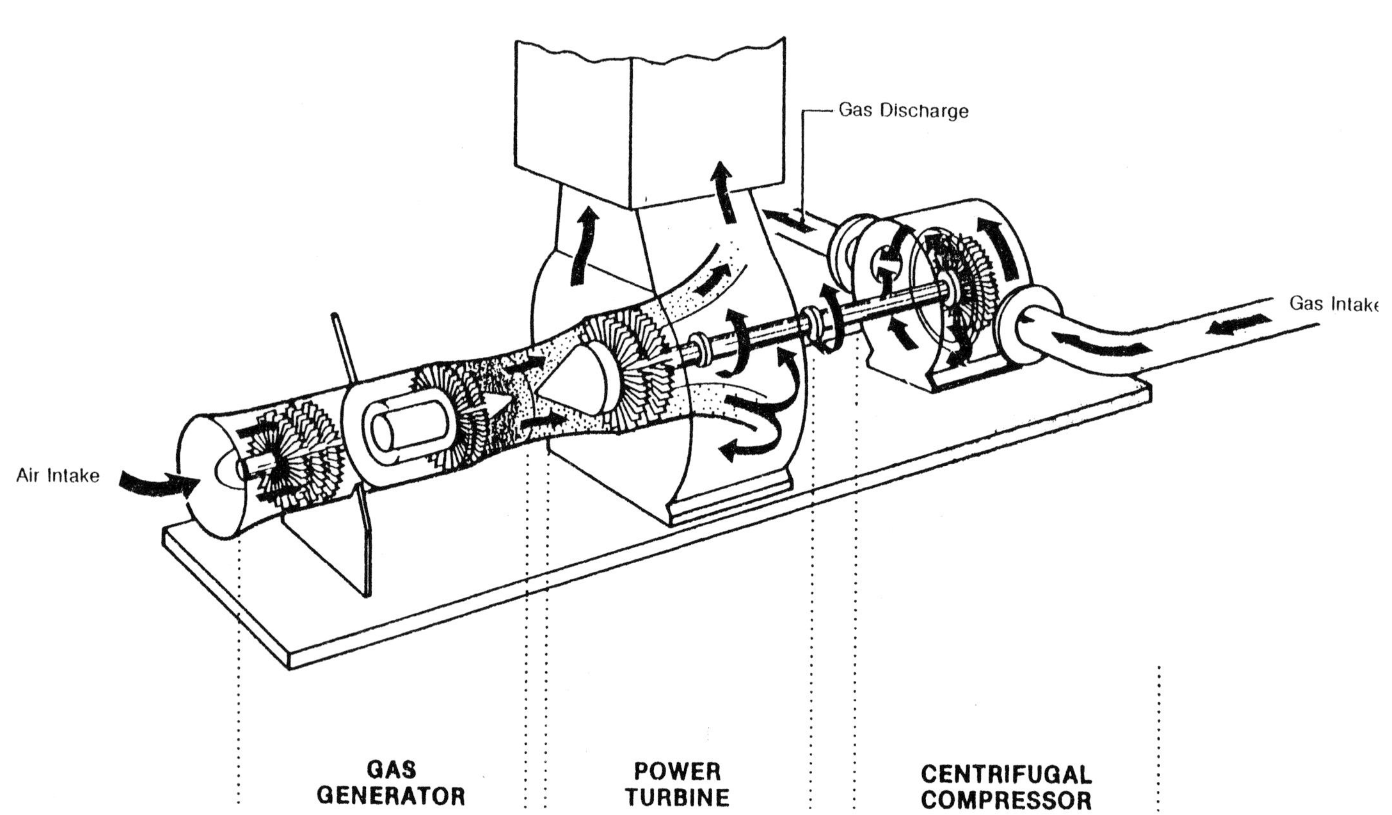

Fig 1 Typical gas turbine compressor ('free' power turbine)

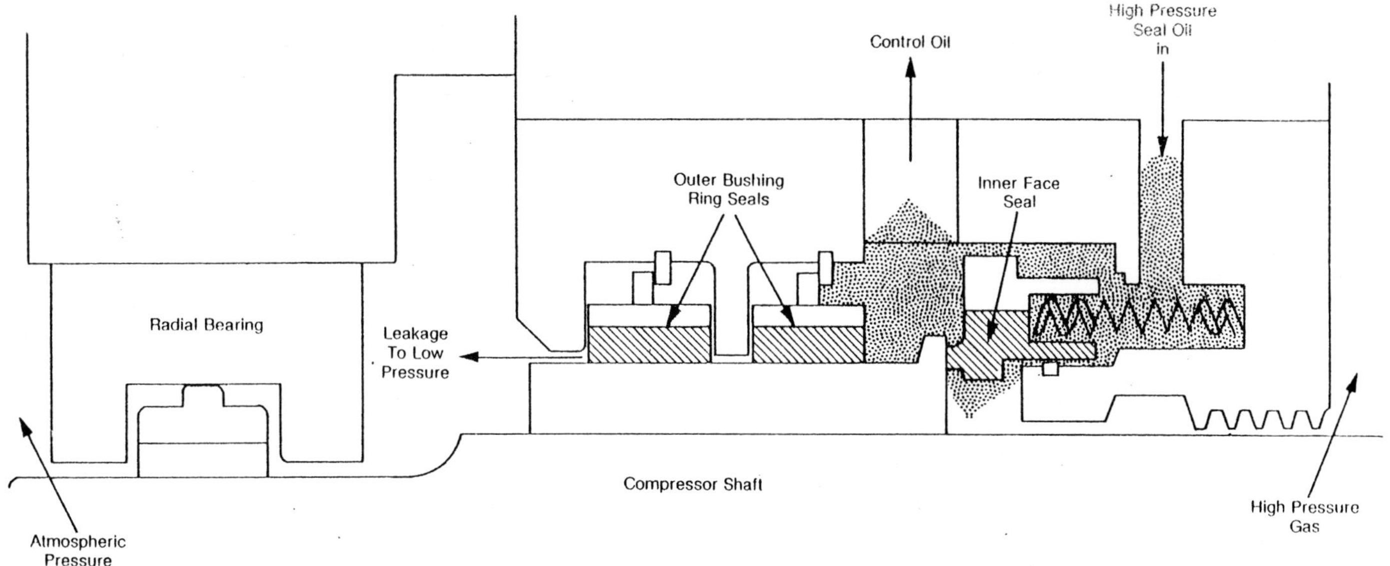

Fig 2 Compressor seal system and bearing arrangement

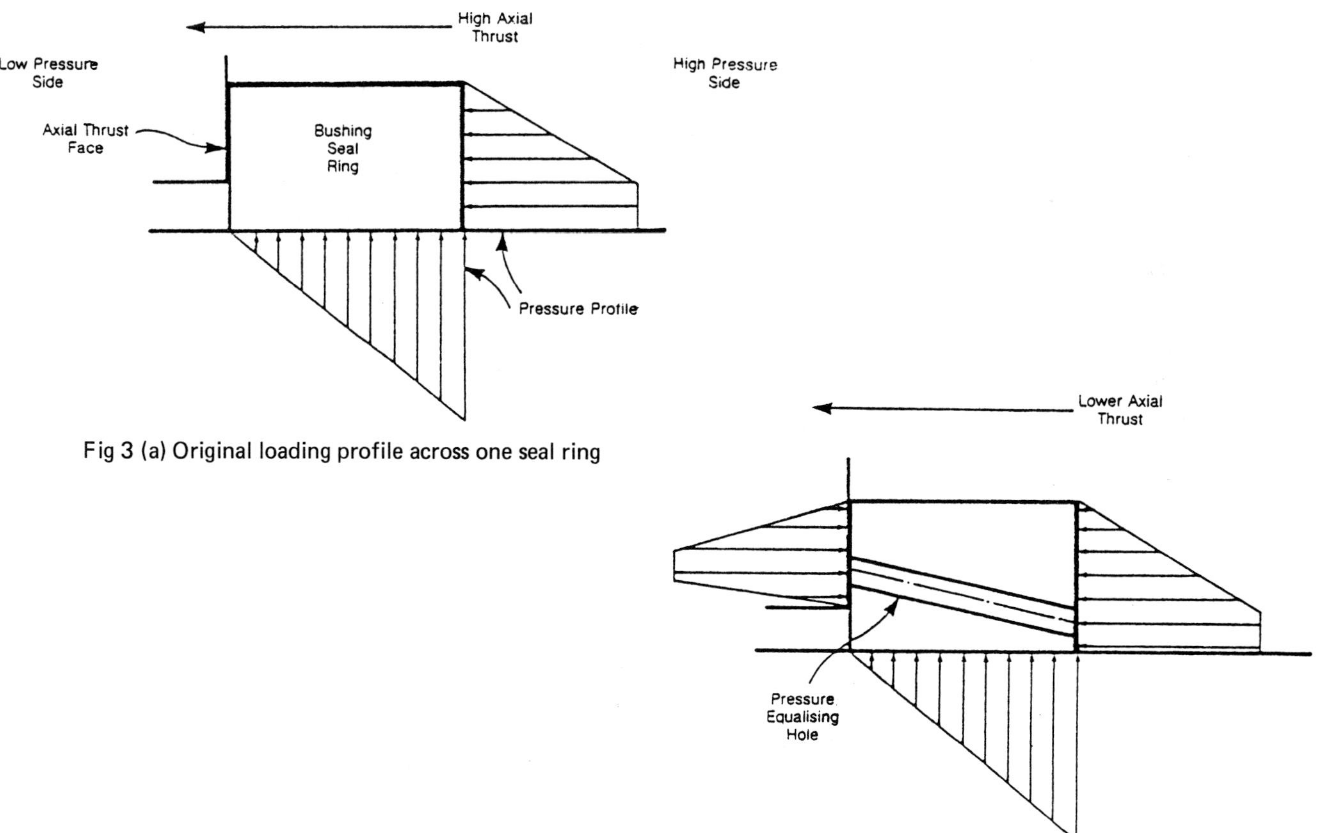

Fig 3 (a) Original loading profile across one seal ring

Fig 3 (b) Modified seal ring giving improved loading profile

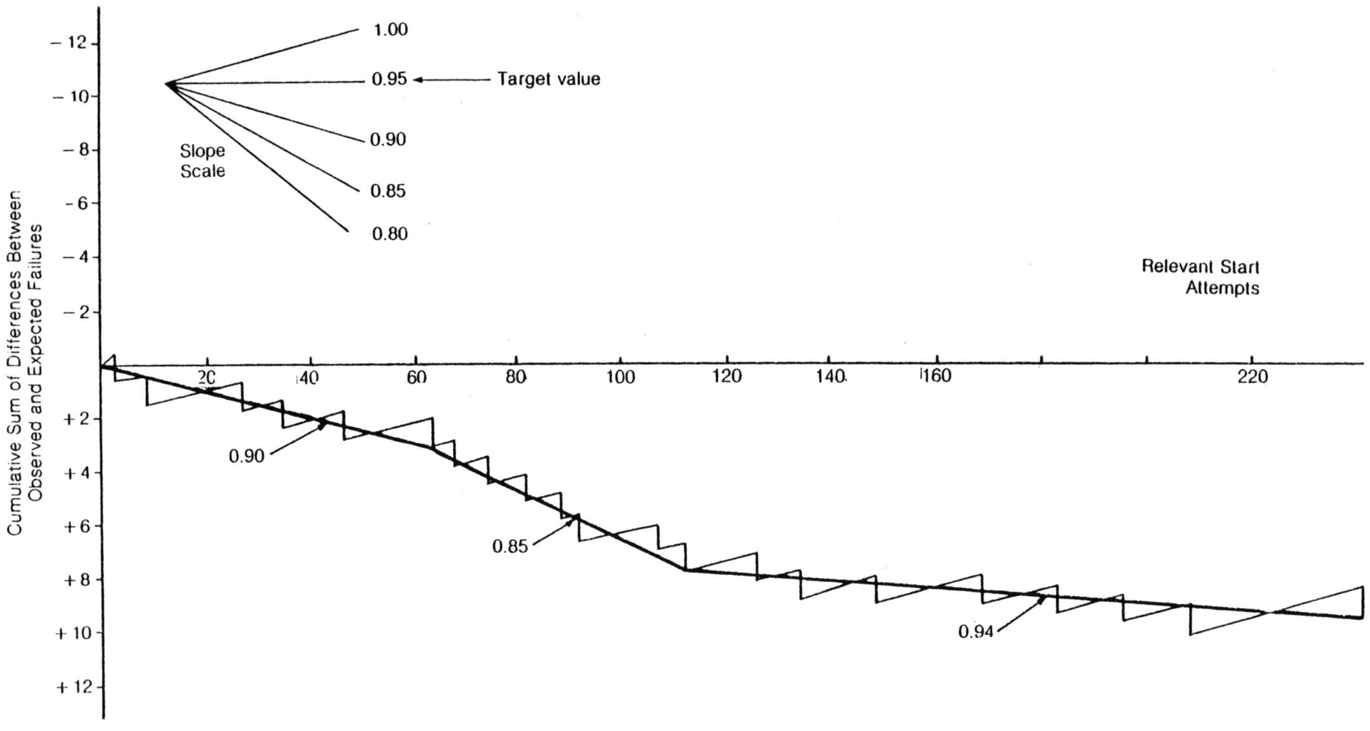

Fig 4 Start reliability cusum plot

Cumulative Sum of Differences Between Observed and Expected Failures
– 4
+ 4
+ 8
+ 12
+ 16
+ 20
+ 24
+ 28
+ 32
+ 36
+ 40
+ 44
+ 48
+ 52
+ 56
Operational Trial Running Hours
200
400
600
800
1000
1200
1400
1600
1800
2000
∞
475
Target Value
200
100
50
25
Slope Scale
6
20
160

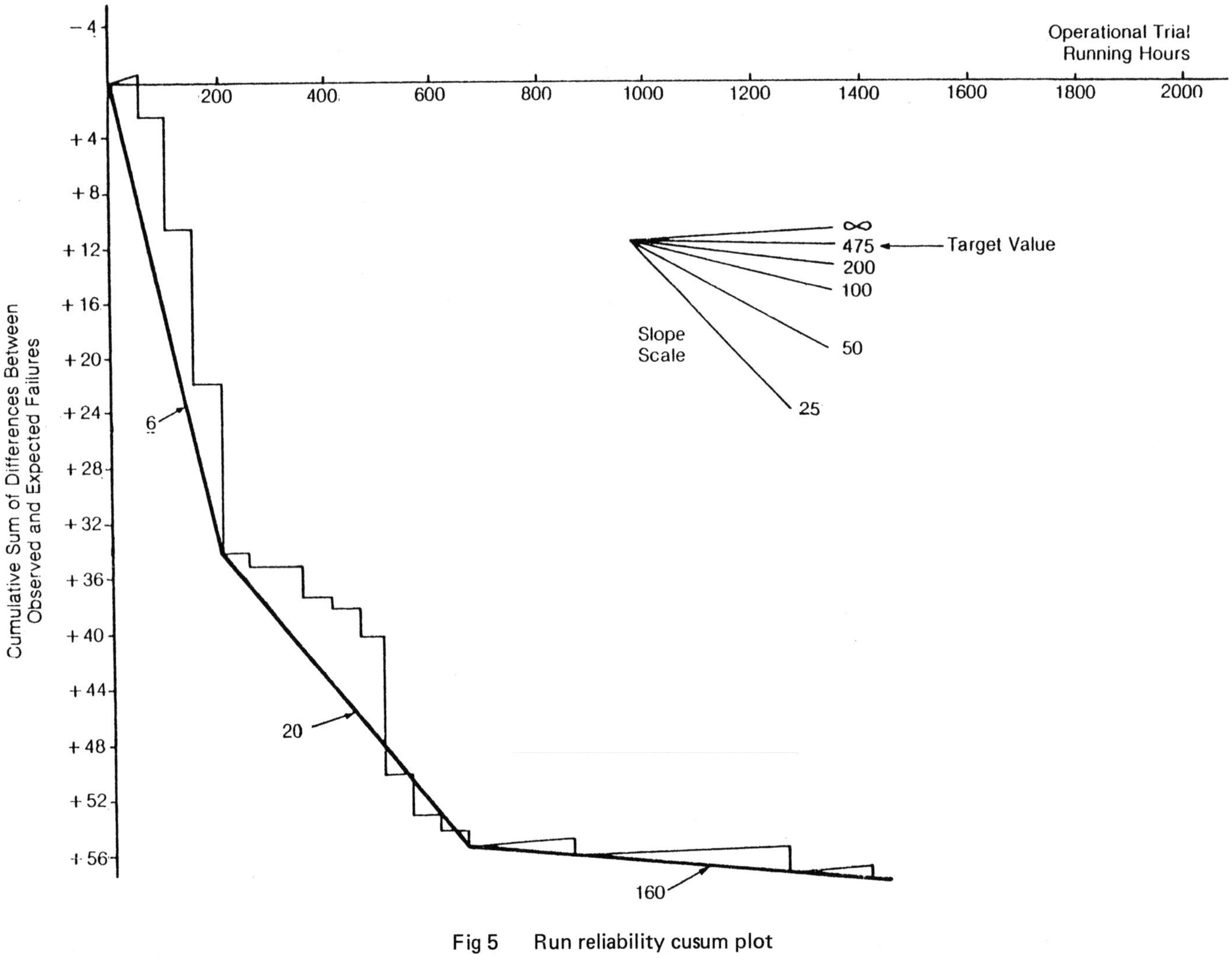

Fig 5 Run reliability cusum plot

Factors affecting reliability specifications for process plant

D J SHERWIN, PhD, MSc, CEng, MIMechE
Department of Engineering Production, University of Birmingham

SYNOPSIS

The importance of reliability specification is explained and advice given with respect to the process industries. Progress towards a new British Standard on reliability specification is reported.

1. INTRODUCTION

For continuous process plants the most commercially important measure in Reliability is usually the overall productiveness or average possible output relative to rated full output. Considerations of high first cost and high interest rates tend to produce plants consisting of a single series of stages with little duplication. Reliability and Maintainability become very important factors in the calculation of Life Cycle Costs (LCC) and expected profits because failures often mean complete cessation of output during repairs. The plant as a whole is less reliable than its most fallible stage.

Process equipment can usually be built to meet any reasonable Reliability specification. Buyers generally get what they pay for provided that they ask for what they want. Plant manufacturers, however, need detailed objective feed-back on the reliability and performance of their products if they are to advance the state of their art to mutual benefit. This should become in the near future, a principal activity of the Engineering Equipment and Materials Users Association (EEMUA) who co-operated in the preparation of this paper. The Institution also could provide facilities such as access to data banks.

Single-stream plants operating to capacity to meet buoyant demand often suffer from temporary or chronic lack of preventive and restorative maintenance. It is important that preventive maintenance (PM) requirements are known and considered at the design stage and included in the specification agreed between maker and user. The effect of PM on plant short-term Reliability and long-term Durability can be considerable.

It is concluded that Reliability specifications for new plant are a good idea, but that they should be drawn up in the light of objective analysis of part failure data and the system availability analysis of the plant as a whole, including PM requirements, and on the basis of assessment of the customer's life-cycle costs.

The BSI is at present pondering a new standard on Reliability Specification. Intelligent specification by users in consultation with manufacturers of plant is a key to the reversal of the present tendency for British users to buy foreign plant. Recognition of the need for Reliability specification will, it is hoped, lead to the provision of the necessary data collection and analysis facilities and encourage manufacturers to adopt a policy in which the customers' costs over the life cycle, including lost production costs are their paramount criterion.

The meanings of specialised Reliability Engineering terms used in this paper are in BS.4778(1), but will in most cases be clear from the context.

2. FORMS OF RELIABILITY SPECIFICATION AND THEIR DRAWBACKS

There are several ways in which the Reliability of any item may be specified:

(a) By specifying on the basis of experience what materials and standards of fabrication should be used, eg, pipework to ASME standards appropriate to the technology.

(b) By specifying a particular model of, for example, a pump on the basis that it has given good service elsewhere under similar conditions or is guaranteed by its manufacturer to do so.

(c) By specifying numerically the acceptable minimum requirements for Reliability and Maintainability or Availability under a specified PM regime given the performance required and a full description of the working environment. For example 'The pump is to have a mean time between failures (MTBF) of not less than 17 000 hours and a mean time to repair (MTTR) of not more than 10 hours with PM allowed every 5000 hours. Notwithstanding the above requirements the intrinsic availability is to be better than 99.98% on a timescale which does not include the PM. The total downtime is not to exceed 40 hours average. A pump shall be deemed to have failed if it cannot deliver X litres/sec at Y bar or leaks more than 0.02 litre/min to atmosphere. The working environment is a North Sea Oil platform in an exposed position as defined in specification Z attached.'

(d) By specifying a test, or series of tests, of combined performance and reliability which the item must pass to be accepted into service. These tests may be at increased stress and/or in environmental cycles more severe than expected in service. For power plant, installation trials would come under this heading.

Method (a) relies upon concensus as to what constitutes an adequate (safety and) Reliability standard. In such cases a lot of people over many years have considered much data and many factors and come to the agreed conclusion that this is the prudent way to engineer such systems. Statistical basis is largely absent but as failures are rare, this does not matter much. The majority of failures are to moving machinery rather than pipes and static vessels. Even using Method (a), PM in the form of periodic inspections, pressure tests, corrosion protection etc. is needed to uphold the assurance of adequate Reliability over a plant life cycle. Life is ultimately limited despite prescribed PM measures being conscientiously performed.

Users of Method (b) are warned that small variations in duty or environment can give rise to order of magnitude changes in average failure rate. This observed phenomenon has been theoretically explained most elegantly by Prof. A. D. S. Carter (4, 5) whose work also shows that where a component proves inadequate in a changed duty or environment, it has only to be strengthened a little to restore the failure rate to an acceptably low level, or (more precisely) to delay all except randomly caused failures for an acceptably long average time. The theory depends upon a model in which both load and strength are distributed and where the strength distribution, due to some form or forms of progressive weakening marches into the load distribution causing more and more of the population to fail as time passes (see Figure 1). The fit to a distribution form is likely to be less good in the tails and so the reliability effects of small movements in mean load are difficult to predict. Neglect or ignorance of these factors has led some plant manufacturers to narrow the margin between load and strength by producing a faster version of an existing design without strengthening the moving parts. For example a motor car engine may reasonably be expected to last 100 000 miles in the ordinary way but the rally version of the same engine is life-expired after each rally, two orders of magnitude sooner in mileage terms. Turning the phenomenon to advantage, diesel engines for Royal Navy service are usually de-rated in power by about 10%. The same engine, at full power in trains, has given much trouble whereas de-rated it has been satisfactory in the Navy.

As an example of the effect of environment, a typical electrical component's failure rate increased by 100 fold for an increase of operating temperature from 20C to 80C. In this case the law of variation is that of Arrhenius, i.e. for electronic components

$$\text{Mean life} = A.\exp(E/kT)$$

where A is a constant, E the activation energy, k is Boltzmann's constant (8.63×10^{-5} ev/K) and T the absolute temperature (K). Other environmental factors such as vibration and humidity have less predictable effects.

The problem with Method (c) is to prove that the buyer has actually got goods which meet the specification without expensive overkill. Assuming a Poisson process for the generation of failures, to demonstrate that the MTBF exceeded 17 000 hours with 95% confidence would require total testing of about 3 times that number of pump-hours and even more if any failures occurred (6 Part 2). Accelerated tests using enhanced environmental conditions and/or mechanical or electrical overload are subject to uncertainties like Method (b).

In Method (d) the user takes responsibility for the method of proving reliability which, in this case, is indirect. Overstress and environmental tests tend to reveal relevant failure syndromes earlier, but they also sometimes induce failures which would not occur in service. Installation trials are necessary to eliminate early failures and to train operators and maintainers.

In the following sections, some suggestions will be made for overcoming the drawbacks and towards a regime which would allow suppliers to sell and buyers to buy with confidence in reliability and performance reliability; that is the objective of specification.

3. THE NEED FOR DATA COLLECTION

A detailed case for data collection from the field was made in a previous paper (7). The main thesis of (7) was the improvement of PM schedules and their cost-effectiveness. Three points are of interest in the present context.

(a) Data collection and analysis is often rejected on grounds of cost. Its marginal cost need not be great if collection is organised as part of an integrated management data system. Some data on plant failure and maintenance has to be collected in order to retain primary financial and operational control and to comply with the minimum requirements of the law in respect of accounts and safety. Critical but unprejudiced examination of the marginal costs usually shows them to be much exceeded by the possible savings which more detailed knowledge of plant behaviour permits. Figures 2 and 3 taken from (7) by kind permission of the I.Plant.E show respectively an outline of a suitably integrated management information system and the growth of utility with detail in reliability data generally.

(b) The collection and analysis of reliability data allows the user to check that specified reliability is being achieved, and to suggest improvements to the manufacturer to improve present equipment.

(c) The greatest benefit is to future plant specifications. Even if the data collected is not passed back to manufacturers, at least the user will be able to avoid some of the specification mistakes of the past. This may be no more than avoiding certain brand-names for particular applications. In other cases, analysis of failure modes will reveal the need for some particular feature or alternatively that an expensive frill such as glass-lining against acid attack is not actually necessary for adequate reliability.

It is unhelpful simply to collect data on how often and expensively an item failed and then to specify for the next order that it must be improved to certain levels. It is helpful to the manufacturer to detail the modes or syndromes of failure that were most common - then he has some chance of improving the engineering of the next generation. He is more likely to have an improved version available when wanted if he has been informed regularly of the problems arising. In the Defence field there is two-way traffic of uniformed engineering personnel between the design authorities and the users. In industry this is less common. An isolated designer never learns to do better because nobody tells him where he is going wrong. Customers suffering failures which they believe are due to design faults should insist on speaking to the suppliers' design section. They will sometimes learn that really they have a specification problem; because the designer never intended his product for the duty imposed. Sales and service engineers tend, unfortunately, not to tell customers they specified the wrong item for fear of losing orders. A talk with the designer before specifying can save much time and money later, particularly if the evidence is available for comparative assessment of expected Reliability by interpolation between conditions for which data exist. Extrapolation is much more chancy.

4. EFFECT OF PREVENTIVE MAINTENANCE

Data collection provides a record of PM as well as of failures and repairs. The record is useful evidence in case of dispute of the extent to which the user has kept faith with the agreed PM schedule. PM has recently acquired a bad name in Industry. Accountants have difficulty regarding maintenance as an investment to prolong the active life of plant; they tend to see it as an expense which can be cut at least in the short term. Operational Research groups point to cases where they have shown savings by deliberately ceasing preventive measures. This may be also because the PM was being done badly in which case the answer is an investment in training of the maintainers and their supervisors. Reference (2) contains three case studies of this general problem all of which tend to show that properly performed PM is economical. They also demonstrate that PM can have a profound effect upon the average rate of occurrence of failures. It follows that plant reliability specifications should, for the protection of the plant manufacturer and user alike, contain or be associated with an agreed schedule of PM.

In theory, PM can reduce the residual failure rate to any required level short of zero - at a price. Even where the lost time due to PM is not important, this is not in practice pursued very far because of the diminishing returns. Figure 4 shows the familiar bath-tub curve of observed failure rate with the effect of various actions, all of which cost money to implement. For any fixed schedule of PM to given standards, the average costs per machine for a population of similar machines can be found empirically and traced as a cumulative curve against time, starting at the purchase cost at time - zero. For a given PM schedule an economic durability or total life exists at the tangent to this curve at the origin because this minimises the average rate of expenditure (8). The costs should be dis-inflated and reduced to NPV. The cost and bath-tub curves are then seen to be interdependent and malleable, by altering the PM schedule the Quality Control procedures and the training of operators and maintainers.

5. EFFECT OF INITIAL QUALITY & RELIABILITY - NEW TECHNOLOGY

For process plant, Quality Control includes installation inspections and acceptance trials. In (9) it was explained that early unreliability can and has rendered whole projects unprofitable by delaying the move into profit for so long that the accumulated investment debts and interest can never be paid out of subsequent profits under discounted cash flow rules. It seems reasonable to make the equipment suppliers bear some of this risk by specifying dates by which items are to be installed and working satisfactorily and penalties for non-compliance. This is particularly important for new technology items and control systems.

New technology appears in process plant for two principal reasons, new processes and increased scale of output. For new processes, many items will be new and their design specification must include provision for testing and time for corrective action. The same applies with less force to standard items used in new contexts. The case of increased scale provides an opportunity to discuss (partial) redundancy.

The capital expenditure on a new process plant is roughly proportional to the dry weight which is, in turn, roughly proportional the the external area of all the pipes and vessels and pumps, etc. The output is proportional, roughly, to the enclosed volume so that cost varies roughly as the two-thirds power of output. (The exponent is often taken to be 0.63 rather than two-thirds.) This 'law' has led to large single-stream plants to keep down capital costs. There is little reliability risk in scaling up pipes, vessels and other static equipment but there have been some bad experiences with rotating machinery. Before specifying a new and untried design, the advantages of fitting more than one machine of lower output but proven reliability should be explored on a LCC basis. The capital cost may be higher but the plant will produce sooner and produce at least part output for a greater proportion of time. There is great security of part output from a plant which is essentially two or more separate lines with cross-connections between stages (10). In this case the capital cost can be spread, so that the second line is paid for by the profits of the first. Output and maintenance can be timed to best advantage in relation to the product demand cycle. The large single-stream plant may still be the best prospect in some cases but the provision of standbys and one or two or more production lines with cross-connections almost always warrants consideration; on an LCC basis these options cannot be rejected without study.

The investment in process plant projects is often large. The use of LCC and reliability calculations to trace the expected cost-income profile over the life of the plant is generally done, but it is less common to back up the choice of configuration and particular equipment with a thorough reliability study as outlined in BS.5760 Part 1(6). A North Sea Oil production platform design from which several units were eventually built was approved with only the most perfunctory reliability study. As a result the platforms were short of electric power and of control air and suffer from the unreliability of gas compressors of advanced but previously untried design. Little oil is lost because oil/gas separation involves few rotating machines but enough gas is lost annually to supply Birmingham. Only the control air problem could be dealt with by redesign, there was no room to fit more generators or compressors.

6. BSI ACTIVITY ON RELIABILITY SPECIFICATION

BS5760 Part 1 - Guide to Reliability Programme Management was published in 1981. The Chairman of the relevant sub-committee (QMS/2/3) and several other members work or have worked in the process industries. The guide generally speaks from a viewpoint consistent with the customer specifying requirements to a manufacturer because where this is not the case, it is generally suggested that the manufacturers' Reliability staff should represent the customers. This standard guide is intentionally general but it is not vague; it lays down a procedure for new product development which takes account of both new technology and evolutionary development. It applies to manufacturers of bespoke and standard products in all branches of engineering. Some 2000 words and a whole page figure (reproduced here as Fig. 5 by kind permission of BSI) out of 11 000 words and 5 pages of figures are devoted to the important topic of specification for reliability. Figure 5 envisages reliability clauses in each specification and manual but the guide also discusses the alternative policy of writing a comprehensive reliability specification covering all aspects of the development of the design and its subsequent use. It emphasises the advantages whilst listing the problems in a quantitative approach. In the qualitative approach the onus lies with the customer - he has judged for example, that adherence to a Code of Practice will produce the required reliability.

In writing reliability clauses the guide advises specification in terms of

a) intended function
b) failure criteria
c) measure of reliability
d) required limiting value of measure
e) environmental conditions
f) 'load' conditions
g) means by which reliability is to be attained and demonstrated.

Despite the fairly heavy emphasis upon specification methodology in Part 1, QMS/2/3 currently have a sub-panel (QMS/2/3/4) investigating a new part of BS5760, presently called Part X. The work has (November 1983) reached the stage of an outline plan and a collection of contributions on some of the topics in the plan. The panel would welcome contributions from the process industries on present practices and why they do or do not work and also examples of reliability specification to be considered for an expanded edition of BS5760, Part 3.

The current draft of Part X expands on the relevant section in Part 1 but also introduces some fresh notions. In following a general policy of QMS/2/3 to include the construction industry, the coverage in respect of process plants has incidentally been improved. Three types of product are categorised: 'Manufactured', 'Constructed', and 'New Systems of Existing items'. It is recognised that the nature of the market will affect who does the specifying. The draft discusses the reliability content of specifications for various stages in the life of an item (see Figure 5). A section is planned on the specification of Reliability Programme management requirements which will be of particular interest to the process industries. The integration of the reliability programme with the overall quality planning is also to be emphasised. Sections on quantitative specification, software aspects and a number of other topics remain to be written.

BS4200, Part 7, (12), is concerned with Reliability Specification but is really only applicable to electronics. It emphasises specification of confidence limits on failure rates which are assumed constant (exponential distribution of failure times). It emanates from the International Electro-technical Commission and is published by BSI to comply with international agreements.

7. CONCLUSIONS

a) Reliability specifications for process plant are a good idea deserving more care and attention than they generally receive at present.

b) The appropriate measure for process plant is usually availability on a time-scale which includes preventive as well as corrective maintenance. This implies
   (i) Trade-off studies between reliability and maintainability and between failure and preventive down-time on the basis of LCC and the profit expenditure profile for the project from inception to disposal.

   (ii) System availability modelling to determine optimum redundancy and output capability versus time.

c) Suppliers and users of plant should establish or support measures to improve the reliability of plant generally by getting the necessary data back to the designers. Specifications should be based on modest improvements on current experience. Numerate specifications are not useful if they have been produced without reference to such data.

d) Reliability requirements should be specified in respect of each stage of the development of the design of a new plant. Whether to use reliability clauses in general specifications for each stage or to have a separate reliability specification covering all stages is a matter of taste.

   The advice in BS5760, Part 1, on specification is applicable to the process industries and it is likely that Part X will be more detailed and more useful.

e) The minimum requirements of a reliability specification are that it should define:

   (i) relevant failure unambiguously in context,

   (ii) the environmental conditions,

   (iii) the level of stress (or some factor from which it may be found),

   (iv) the minimum requirements for reliability, either numerately or by reference to a standard or model number,

   (v) the maintenance policy and, if appropriate,

   (vi) the confidence level at which the minimum requirement is to be proved,

   (vii) the tests to be applied to assure compliance,

   (viii) the envelope of limits for trade-off between the factors affecting LCC.

References

(1) BS4778 Glossary of terms used in quality assurance (including reliability and maintainability terms) British Standards Institution 1981.

(2) SHERWIN D J and LEES F P 'An investigation of the application of failure data analysis to decision-making in maintenance of process plants.' Proc.Instn.Mech. Engrs 1980, 194, 29, pp 301-319.

(3) SHERWIN D J 'Failure and maintenance data analysis at a petrochemical plant.' Reliability Engineering 1983, 5 pp 197-215.

(4) CARTER A D S 'The bath-tub curve for mechanical components - fact or fiction?' Instn.Mech.Engrs' Conf. on Improvement of Reliability in Engineering, Loughborough, Paper No. C75/73. 1973.

(5) CARTER A D S 'Achieving quality and reliability' (1974) (James Clayton lecture) Proc.Instn.Mech.Engrs. 188,13.

(6) BS.5760 Reliability of Systems equipments and components, Part 1 (1979), Guide to reliability programme management Part 2 (1981), Guide to the assessment of reliability, Part 3 (1983), Examples. (Further part X projected on specification of reliability.) British Standards Institution, London.

(7) SHERWIN D J 'Reliability data for better maintenance management. Instn Plant Engrs Symposium, Falkirk. 1980

(8) DRINKWATER R W and HASTINGS N A J, 'An economic replacement model' Operational Research Q'ly 1967 18, 2, pp 121-138.

(9) HOLROYD, Sir Ronald, 'Ultra-large single-stream chemical plant; their advantages and disadvantages'. Presidential address to Society of Chemical Industry published in 'Chemistry and Industry' 1967, August pp 1310-1315.

(10) SHERWIN D J, 'Availability and productiveness of systems of redundant stages of like items in terms of item availability'. International Journal of Production Research 1984 22, 1, pp 109-120.

(11) MIL-HBK-217C Reliability Prediction of electronic equipment Rome Air Develop-Center, Griffiss Air Force Base NY13441 USA (1979)

(12) BS4200 Reliability of electronic systems and components Part 7 - Specification British Standards Institution 1983.

pdf
time
Load
Strength

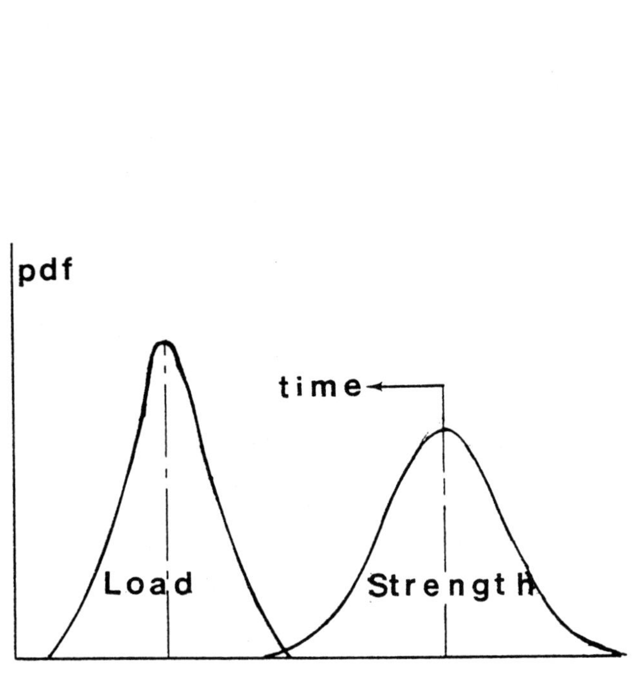

Fig 1 Mechanical failure in a population of components is due to overlap of load and strength distributions. Rate of occurrence increases if mean strength deteriorates with time (after Carter Refs 4, 5)

Job cards
Vouchers
Invoices
Work measurement
Stores system
Contracts system
Reports
Engineering cost system
Engineer inspection reports
Reports
Equipment history system
Main finance system
Reports and microfiche
Fixed costs system
Profit and loss accounts
Reports

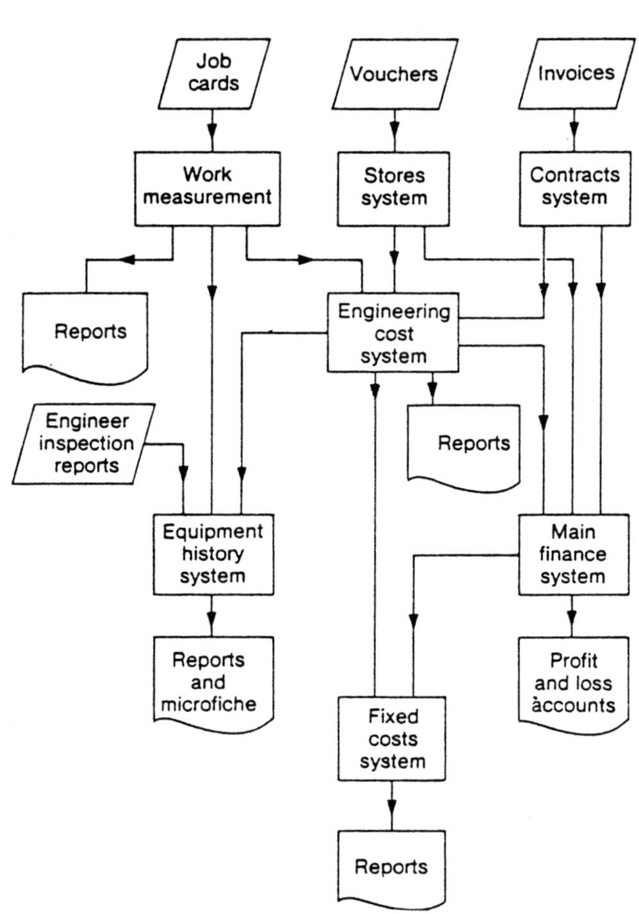

Fig 2 Outline of an integrated management information system (reproduced from Ref 3 by kind permission of the editor 'Reliability Engineering)

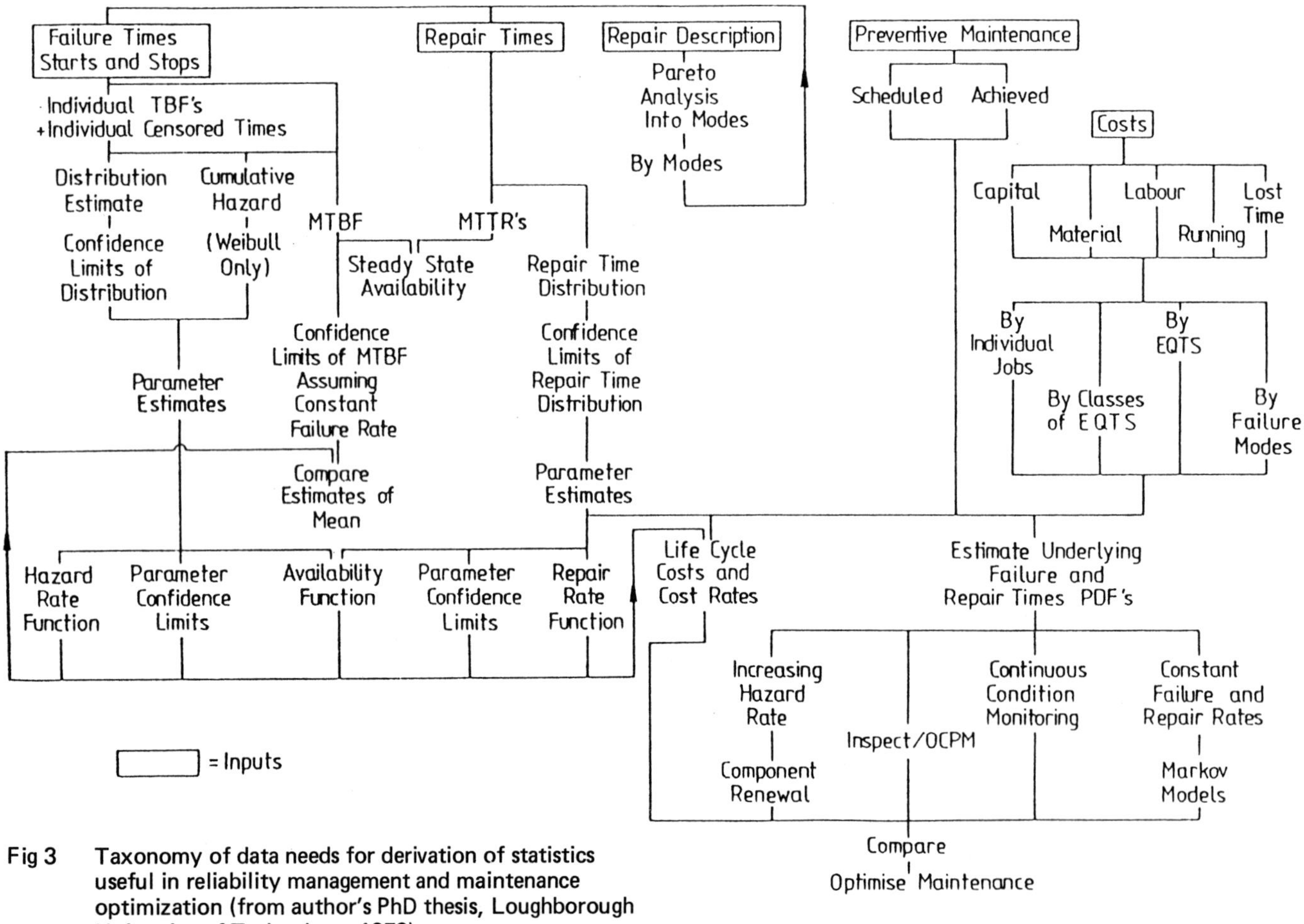

Failure Times Starts and Stops
Repair Times
Repair Description
Preventive Maintenance
Individual TBF's
+Individual Censored Times
Pareto Analysis Into Modes
By Modes
Scheduled
Achieved
Costs
Distribution Estimate
Cumulative Hazard
Confidence Limits of Distribution
(Weibull Only)
MTBF
MTTR's
Steady State Availability
Repair Time Distribution
Capital
Labour
Lost Time
Material
Running
Confidence Limits of MTBF Assuming Constant Failure Rate
Confidence Limits of Repair Time Distribution
By Individual Jobs
By EQTS
By Classes of EQTS
By Failure Modes
Parameter Estimates
Compare Estimates of Mean
Parameter Estimates
Hazard Rate Function
Parameter Confidence Limits
Availability Function
Parameter Confidence Limits
Repair Rate Function
Life Cycle Costs and Cost Rates
Estimate Underlying Failure and Repair Times PDF's
Increasing Hazard Rate
Continuous Condition Monitoring
Constant Failure and Repair Rates
Inspect/OCPM
Component Renewal
Markov Models
Compare
Optimise Maintenance
= Inputs

Fig 3 Taxonomy of data needs for derivation of statistics useful in reliability management and maintenance optimization (from author's PhD thesis, Loughborough University of Technology, 1979)

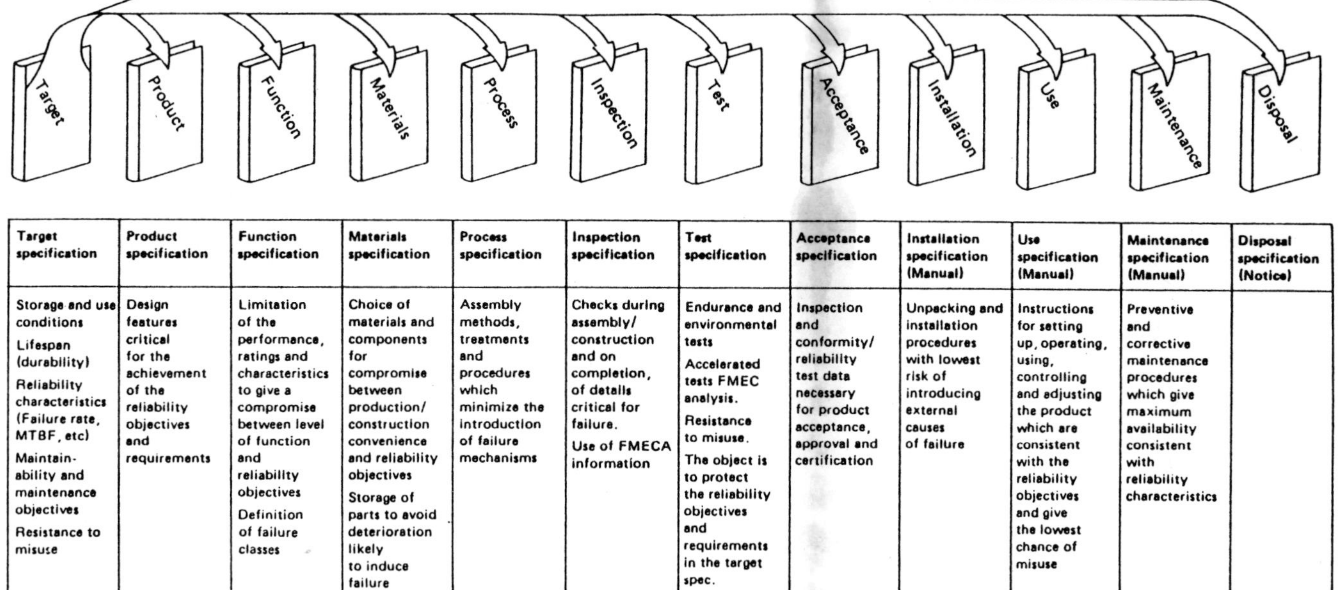

Target specification	Product specification	Function specification	Materials specification	Process specification	Inspection specification	Test specification	Acceptance specification	Installation specification (Manual)	Use specification (Manual)	Maintenance specification (Manual)	Disposal specification (Notice)
Storage and use conditions Lifespan (durability) Reliability characteristics (Failure rate, MTBF, etc) Maintainability and maintenance objectives Resistance to misuse	Design features critical for the achievement of the reliability objectives and requirements	Limitation of the performance, ratings and characteristics to give a compromise between level of function and reliability objectives Definition of failure classes	Choice of materials and components for compromise between production/ construction convenience and reliability objectives Storage of parts to avoid deterioration likely to induce failure	Assembly methods, treatments and procedures which minimize the introduction of failure mechanisms	Checks during assembly/ construction and on completion, of details critical for failure. Use of FMECA information	Endurance and environmental tests Accelerated tests FMEC analysis. Resistance to misuse. The object is to protect the reliability objectives and requirements in the target spec.	Inspection and conformity/ reliability test data necessary for product acceptance, approval and certification	Unpacking and installation procedures with lowest risk of introducing external causes of failure	Instructions for setting up, operating, using, controlling and adjusting the product which are consistent with the reliability objectives and give the lowest chance of misuse	Preventive and corrective maintenance procedures which give maximum availability consistent with reliability characteristics	

Fig 4 'Bath-tub' and total cost curves for a complex equipment showing the effects of maintenance, training and quality control on overall rate of expenditure and economic life

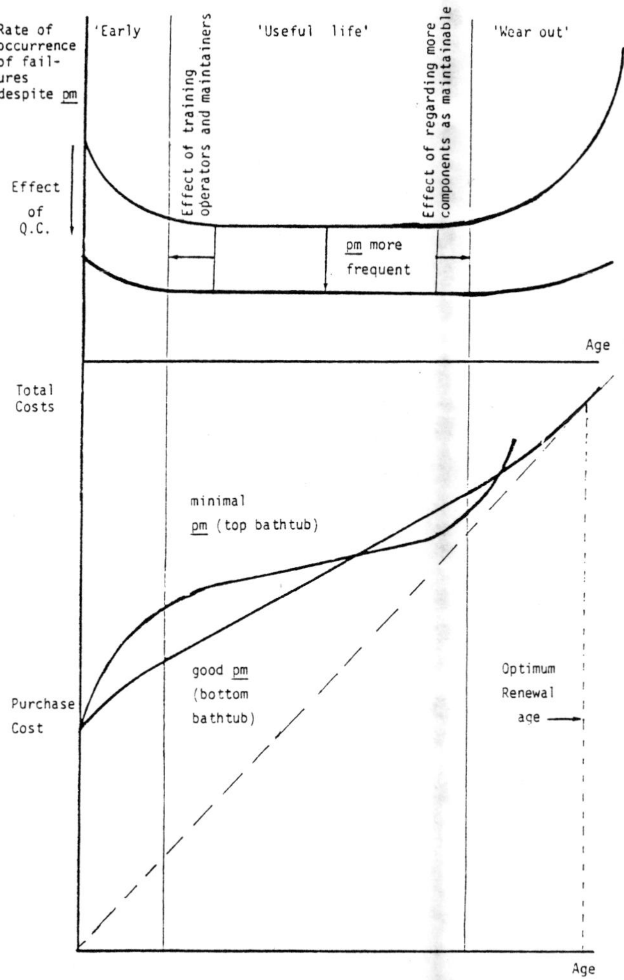

Fig 5 'The reliability content in specifications for manufactured and constructed products', reproduced from BS.5760 Part 1, by kind permission of the British Standards Institution